BAD AIR
WHAT ARE WE BREATHING?

Kevin B DiBacco

CHELSEA HOUSE PRESS

CONTENTS

THE INVISIBLE THREAT AROUND US

Take a deep breath. Go ahead, I'll wait.

Did you think about what just entered your lungs? Probably not. Most of us don't. We breathe about 20,000 times a day, pulling in roughly 2,000 gallons of air, and we barely give it a second thought. That's totally normal – breathing is just something we do, like blinking or walking. But here's the thing: maybe we should be paying more attention to what's floating around in that air we're gulping down all day long.

Let's talk about what's really happening with our air in 2024. We're facing some unprecedented challenges:

Urban Air Quality Crisis:

- Megacities struggling with smog
- Traffic pollution reaching new heights
- Construction dust everywhere
- Industrial emissions mixing with urban life
- Chemical cocktails from modern living

Think about your typical morning commute. That twenty-minute drive to work? You're not just traveling through space – you're traveling through a complex soup of:

- Vehicle exhaust from thousands of cars
- Construction dust from that new high-rise
- Industrial emissions from nearby factories
- Ground-level ozone forming in the sunlight
- Particulate matter from brake dust and tire wear

I remember the first time I really thought about air quality. I was stuck in traffic on a sizzling summer day, windows down because my car's AC was on the fritz. The truck in front of me belched out a cloud of black diesel exhaust, and I could actually taste it. Gross, right? But that got me thinking – if I could taste and smell that exhaust, what about all the stuff in the air I couldn't detect? What was I breathing in every single day without even knowing it?

Modern Air Challenges:

Chemical proliferation:

    Thousands of new compounds yearly
    Unknown long-term effects
    Complex interactions
    Synthetic materials everywhere
    Industrial innovations

Climate change impacts:

    Increased wildfire smoke
    Rising ground-level ozone
    Changed weather patterns
    Extended pollen seasons

Heat-pollution interactions

The Indoor Air Crisis:

Building materials:
- Off-gassing furniture
- Synthetic carpets
- Paint and adhesives
- Pressed wood products
- Flame retardants

Modern lifestyle impacts:
- Work-from-home pollution
- Electronic device emissions
- Cleaning product chemicals
- Personal care products
- Cooking fumes

You know what's crazy? That nasty truck exhaust moment was actually a gift in disguise. It opened my eyes to something we all take for granted. See, most air pollution is invisible. You can't see it, smell it, or taste it. It's like having a party crasher in your house that you don't even know is there.

Let's break down what's actually in our modern air:

Outdoor Pollutants:

Vehicle emissions:
- Carbon monoxide
- Nitrogen oxides
- Particulate matter

Volatile organic compounds
Heavy metals
Industrial contributions:
Sulfur dioxide
Industrial solvents
Metal particles
Process emissions
Chemical releases

Indoor Pollutants:

Building-related:
Formaldehyde
VOCs from materials
Radon
Asbestos
Lead dust
Activity-generated:
Cooking particles
Cleaning chemicals
Personal care products
Hobby materials
Pet dander

Emerging Concerns:

Microplastics in air:
Synthetic fiber breakdown
Tire wear particles
Industrial processes

Packaging degradation

Textile wear

Nanoparticles:

Engineered materials

Consumer products

Industrial processes

Vehicle emissions

Electronic components

Here's a disturbing thought about what's in your average breath today:

Natural Components:

Traditional elements:

Nitrogen

Oxygen

Argon

Carbon dioxide

Water vapor

Natural particles:

Pollen

Dust

Spores

Sea spray

Volcanic particles

Human-Added Elements:

Chemical compounds:

Industrial emissions

Vehicle exhaust
Consumer products
Building materials
Agricultural chemicals
Synthetic materials:
Microfibers
Plastic particles
Flame retardants
Artificial fragrances
Engineered nanoparticles

The Health Stakes:

Immediate impacts:
Respiratory irritation
Allergic reactions
Headaches
Eye irritation
Throat discomfort
Long-term concerns:
Respiratory diseases
Cardiovascular problems
Cancer risks
Developmental issues
Neurological effects

Environmental Justice:

Unequal exposure:
Low-income communities

Minority neighborhoods

Industrial zones

Traffic corridors

Urban heat islands

Access disparities:

Healthcare resources

Air quality information

Protection measures

Clean air zones

Political voice

Solutions Landscape:

Personal actions:

Air quality monitoring

Filtration systems

Ventilation improvements

Product choices

Lifestyle adaptations

Community efforts:

Policy advocacy

Education initiatives

Monitoring networks

Green spaces

Clean transportation

The Way Forward:

Technology solutions:

Better monitoring

Improved filtration
Clean energy
Smart buildings
Electric vehicles
Policy needs:
Stronger regulations
Environmental justice
Clean air standards
Enforcement
International cooperation

Throughout this book, we'll explore each of these aspects in detail, always focusing on practical solutions and actionable steps. Because while the air quality situation might seem overwhelming, understanding is the first step toward improvement.

Remember, every breath matters. And by the end of this book, you'll understand exactly what's in that breath, why it matters, and most importantly, what you can do to make it cleaner and healthier for yourself and everyone around you.

Ready to dive in? Let's explore what's really in the air we breathe, one breath at a time.

Let's take a look into what's really happening with our air in 2024. We're facing some unprecedented challenges that would have been hard to imagine even a few decades ago:

Urban Air Quality Crisis:

Megacities struggling with smog:
- Photochemical smog formation:
    - NOx and VOC interactions
    - Temperature inversions
    - Trapped pollution layers
    - Peak hour concentrations
    - Weather pattern impacts
- Building density effects:
    - Street canyon pollution
    - Reduced air circulation
    - Heat island intensification
    - Ventilation blockage
    - Pollution trapping

Traffic pollution reaching new heights:
- Vehicle emission types:
    - Primary exhaust gases
    - Secondary formation products
    - Brake and tire wear
    - Road dust resuspension
    - Fuel evaporation
- Traffic pattern impacts:
    - Rush hour peaks
    - Idle emissions
    - Stop-and-go effects
    - Highway corridors
    - Intersection hotspots

Construction boom impacts:

Dust sources:
- Excavation activities
- Material handling
- Equipment emissions
- Concrete cutting
- Demolition work

Chemical releases:
- Solvent evaporation
- Coating emissions
- Adhesive off-gassing
- Waterproofing compounds
- Treatment chemicals

Industrial emissions complexity:

Stack emissions:
- Combustion products
- Process gases
- Particulate matter
- Heavy metals
- Organic compounds

Fugitive emissions:
- Equipment leaks
- Storage tank vapors
- Loading operations
- Waste treatment
- Material handling

Modern chemical cocktails:

Consumer product emissions:
- Personal care products

Cleaning supplies

Air fresheners

Electronics off-gassing

Furniture treatments

Building material releases:

New construction materials

Renovation products

Interior finishes

Insulation compounds

Waterproofing agents

Now, let's break down exactly what's in our modern air, pollutant by pollutant:

Criteria Pollutants:

Particulate Matter (PM):

PM2.5 (Fine particles):

Composition variations:

Metals

Carbon compounds

Sulfates

Nitrates

Organic matter

Source specifics:

Combustion processes

Chemical reactions

Industrial processes

Vehicle emissions

Agricultural activities

PM10 (Coarse particles):

Material types:

Dust

Pollen

Mold spores

Sea salt

Tire wear

Generation sources:

Construction

Road wear

Natural erosion

Agricultural operations

Industrial processes

Gaseous Pollutants:

Ground-level Ozone (O3):

Formation mechanisms:

NOx and VOC reactions

Sunlight interaction

Temperature dependence

Wind patterns

Seasonal variations

Peak conditions:

Summer afternoons

High traffic periods

Stagnant air

Urban heat islands

Regional transport

Nitrogen Oxides (NOx):

Species breakdown:

- Nitrogen dioxide (NO2)
- Nitric oxide (NO)
- Other nitrogen compounds
- Reaction products
- Secondary formations

Major sources:

- Vehicle exhaust
- Power plants
- Industrial processes
- Home heating
- Agricultural burning

Sulfur Dioxide (SO2):

Emission characteristics:

- Industrial releases
- Power generation
- Marine vessel exhaust
- Metal processing
- Coal burning

Transformation processes:

- Acid rain formation
- Particle creation
- Chemical reactions
- Deposition patterns
- Atmospheric residence

Carbon Monoxide (CO):

Source profiles:

- Vehicle exhaust

Indoor combustion

Industrial processes

Natural sources

Waste treatment

Exposure patterns:

Traffic hotspots

Indoor accumulation

Seasonal variations

Urban concentrations

Microenvironment levels

Emerging Pollutants:

Ultrafine Particles:

Characteristics:

Size < 0.1 micrometers

High number concentration

Large surface area

Deep lung penetration

Blood-brain barrier crossing

Sources:

Combustion processes

Vehicle emissions

Industrial activities

Consumer products

Latest technologies

Novel Organic Compounds:

PFAS (Forever Chemicals):

Consumer products

Industrial processes
Firefighting foam
Water-resistant materials
Food packaging
Emerging VOCs:
New building materials
Advanced electronics
Modern furnishings
Personal care products
Cleaning innovations

Air Toxics:

Industrial Chemicals:
Benzene:
Gasoline evaporation
Industrial processes
Consumer products
Tobacco smoke
Indoor sources
Formaldehyde:
Building materials
Furniture
Consumer products
Combustion processes
Natural sources
Heavy Metals:
Industrial emissions
Waste incineration

> Mining activities
>
> Vehicle wear
>
> Historical contamination

Indoor Air Specifics:

Building-Related:

> Construction Materials:
>
>> Pressed wood
>>
>> Insulation
>>
>> Paints
>>
>> Adhesives
>>
>> Sealants
>
> Building Systems:
>
>> HVAC emissions
>>
>> Ductwork contamination
>>
>> Ventilation issues
>>
>> System maintenance
>>
>> Operation patterns

Occupant-Generated:

> Human Activities:
>
>> Cooking emissions
>>
>> Cleaning products
>>
>> Personal care items
>>
>> Hobbies and crafts
>>
>> Electronic device use
>
> Biological Sources:
>
>> Human bioeffluents
>>
>> Pet dander

Dust mites

Mold growth

Bacterial emissions

The complexity of modern air pollution isn't just about the individual pollutants – it's about how they interact and transform in our atmosphere. Think of it as a giant chemistry experiment happening right over our heads, with new compounds being added to the mix almost daily.

Let's shine a light on the invisible troublemakers lurking in our homes – the indoor air pollutants that most people never even think about. It's like having uninvited guests at a party, except these guests are microscopic and potentially harmful:

Hidden Chemical Emissions:

Electronics and Office Equipment:

Computer heat emissions:

Flame retardant releases

Plastic off-gassing

Circuit board compounds

Dust combustion products

Electronic component degradation

Printer/copier emissions:

Toner particles

Ozone generation

Paper dust

Volatile organic compounds

Heat-induced releases

Charging cables/power strips:

- PVC degradation products
- Heat-related emissions
- Insulation breakdown
- Metal particulates
- Coating compounds

Silent Bathroom Pollutants:

Hidden sources:

- Shower steam effects:
    - Chlorine gas release from hot water
    - Mold spore distribution
    - Chemical vapor mobilization
    - Personal care product aerosolization
    - Humidity-activated off-gassing
- Toilet plume aerosols:
    - Bacterial dispersion
    - Viral particles
    - Chemical cleaners
    - Deodorizer chemicals
    - Moisture-carried contaminants
- Medicine cabinet releases:
    - Pharmaceutical dust
    - Personal care product vapors
    - Nail polish remover fumes
    - Hair product residues
    - Expired medication degradation

Kitchen Stealth Pollutants:

Cooking-related:

Non-stick cookware emissions:

Polymer fume fever compounds

Heated coating releases

Metal particle generation

Temperature-dependent toxins

Degradation products

Heated oil aerosols:

Acrolein formation

Fine oil particles

Burnt fat compounds

Smoke constituents

Heat-altered molecules

Hidden appliance emissions:

Refrigerator coil dust

Dishwasher steam chemicals

Microwave radiation effects

Toaster particle generation

Coffee maker volatiles

Bedroom Secret Sources:

Sleeping environment:

Mattress emissions:

Flame retardant off-gassing

Memory foam chemicals

Fabric treatment releases

Dust mite allergens

Moisture-related compounds
Bedding concerns:
Fabric softener residues
Detergent chemical releases
Dryer sheet compounds
Synthetic fiber particles
Treatment chemical vapors
Closet contamination:
Dry cleaning chemical residues
Moth ball vapors
Clothing treatment emissions
Shoe and leather off-gassing
Storage container releases

Hidden Building Structure Issues:

Behind-the-walls problems:
Insulation releases:
Fiberglass particles
Foam insulation gases
Mineral wool fibers
Old asbestos materials
Chemical binder emissions
Wall cavity issues:
Hidden mold growth
Rodent waste particles
Dead insect matter
Wire coating degradation
Pipe corrosion products

Foundation concerns:

- Radon gas seepage
- Moisture-related gases
- Concrete dust
- Efflorescence particles
- Pest treatment residues

Ventilation System Secrets:

HVAC hidden hazards:

- Ductwork contents:
  - Accumulated dust
  - Microbial growth
  - Pest residues
  - Deteriorating insulation
  - Previous renovation debris
- System components:
  - Filter fiber shedding
  - Cooling coil biofilm
  - Condensation pans microbes
  - Motor brush dust
  - Lubricant vapors

Furniture and Decor Emissions:

Unseen releases:

- Upholstered furniture:
  - Foam degradation products
  - Fabric treatment chemicals
  - Stain resistant compounds

Adhesive emissions

Frame material off-gassing

Decorative items:

Artificial plant treatments

Picture frame preservatives

Ceramic glaze particles

Candle soot accumulation

Art supply emissions

Modern Lifestyle Contributors:

Technology and convenience:

Smart home devices:

Battery off-gassing

Electronic component emissions

Plastic housing degradation

Heat-related releases

Wi-Fi router dust attraction

Convenience products:

Plugin air freshener chemicals

Automatic spray dispersants

Reed diffuser evaporation

Scented product emissions

Battery charger heat effects

Pet-Related Hidden Pollutants:

Animal contributions:

Pet care products:

Flea treatment chemicals

Grooming product residues
Litter box dust
Pet bed treatments
Food particle dust
Animal-related particles:
Dried saliva proteins
Skin cell shedding
Fur and dander
Tracked litter dust
Cleaning product residues

Hobby and Leisure Activities:

Hidden activity emissions:
Craft supplies:
Adhesive vapors
Paint and marker fumes
Paper and fabric dust
Plastic craft emissions
Tool operation particles
Exercise equipment:
Rubber mats off-gassing
Equipment lubricant vapors
Foam pad degradation
Plastic equipment emissions
Sweat-related bioaerosols

These hidden pollutants often interact with each other, creating even more complex air quality issues:

Compound Effects:

Chemical interactions:
- Cleaning product mixing:
    - Secondary compound formation
    - Chemical reaction products
    - pH-dependent releases
    - Temperature effects
    - Humidity influences
- Environmental factors:
    - Sunlight-induced changes
    - Temperature variations
    - Humidity impacts
    - Air pressure effects
    - Seasonal variations

The scary part about these hidden pollutants is that their effects often accumulate over time. It's like having a savings account in reverse – instead of building up something good, you're accumulating potential health impacts you might not notice until years later.

Understanding these hidden sources is the first step toward better indoor air quality. In the following chapters, we'll explore specific strategies for identifying and addressing these invisible threats, but for now, just being aware of their existence can help you make better choices about your indoor environment.

Remember, what you can't see can hurt you – but knowledge and proper management can help protect you and your family from these hidden air quality threats.

CHAPTER 1

THE AIR WE BREATHE

Take a moment to notice your next breath. Really focus on it. Feel the air flowing in through your nostrils, filling your lungs, then flowing back out again. This simple act, something you do roughly 20,000 times a day, connects you to one of Earth's most remarkable features – our atmosphere. Have you ever wondered why that breath of fresh mountain air feels so different from the air in a crowded city? Or what's really happening each time your lungs fill with Earth's life-giving mixture of gases? Let's embark on a journey to understand the invisible ocean of air that surrounds us every moment of our lives.

Nature's Perfect Recipe

Think of Earth's atmosphere as nature's most carefully crafted recipe, perfected over billions of years of planetary evolution. Just as a master chef might spend years refining the perfect soup stock, our planet has spent eons developing the exact mixture of gases that makes life as we know it possible. The main ingredients in this atmospheric recipe might surprise you with their proportions:

- **Nitrogen:** A whopping 78% (the base of our atmospheric "broth")
- **Oxygen:** 21% (the essential ingredient for most life)
- **Argon:** Just under 1% (our atmospheric "seasoning")
- **Carbon dioxide:** About 0.04% (a crucial ingredient that requires perfect balance)
- **Trace gases:** tiny amounts of neon, helium, methane, and others (the "finishing touches")

This mixture isn't random – each component plays a vital role in making Earth habitable. You might wonder why nitrogen dominates when oxygen is what we need to breathe. Think of nitrogen as nature's stabilizer, like the flour that thickens a sauce or the egg that binds a cake. It provides crucial atmospheric pressure and serves as an essential nutrient reservoir for life on Earth, even though most organisms can't use it directly from the air.

The Nitrogen Story

The abundance of nitrogen in our atmosphere tells a fascinating story of biological evolution and chemical cycling. While nitrogen makes up the majority of our atmosphere, it exists primarily as N_2 – two nitrogen atoms bound together so tightly that most living things can't break them apart to use them. It's like having a massive pantry filled with canned goods but no can opener.

Nature, however, has developed several ingenious ways to make this atmospheric nitrogen accessible to living things. Deep in the soil, specialized bacteria perform what scientists call "nitrogen fixation" – they break apart those tough nitrogen molecules and combine them with other elements to create compounds that plants can use. Some plants, particularly legumes like peas, beans, and clover, have evolved special relationships with these bacteria, housing them in nodules on their roots in exchange for a steady supply of usable nitrogen.

This bacterial process is so important that without it, Earth's ecosystems would collapse despite being surrounded by nitrogen. Modern agriculture has found ways to artificially fix nitrogen through industrial processes, but these methods consume massive amounts of energy and contribute to various environmental challenges. Nature's bacterial helpers still do the heavy lifting in natural ecosystems, converting about 90% of all fixed nitrogen on Earth.

The Oxygen Revolution

If nitrogen is our atmosphere's foundation, oxygen is its most dynamic player. Each breath you take connects you to one of the most remarkable stories in Earth's history – the rise of oxygen in our atmosphere. The oxygen molecules you just inhaled are participants in a planetary cycle that's been running for billions of years.

Every oxygen molecule in the air has likely been recycled countless times through countless organisms. The oxygen atom that just entered your bloodstream might once have been part of a Tyrannosaurus rex's breath, or perhaps it was released by an ancient redwood tree, or maybe it bubbled up from photosynthetic bacteria in a prehistoric ocean.

But oxygen wasn't always part of Earth's atmosphere. About 2.5 billion years ago, our planet's air was almost completely devoid of this now-essential gas. The revolution began with tiny organisms called cyanobacteria – the world's first oxygen-producing photosynthetic life forms. These microscopic pioneers developed the ability to split water molecules using sunlight's energy, releasing oxygen as a waste product.

This process, which we now call photosynthesis, gradually transformed Earth's atmosphere in what scientists call the Great Oxidation Event. But this wasn't a peaceful transition. The oxygen produced by cyanobacteria was toxic to most of the life forms that existed at the time. Imagine the air itself becoming poisonous – that's exactly what happened to Earth's early life forms. This massive change in atmospheric composition led to one of the largest extinction events in Earth's history, demonstrating the profound impact that air quality can have on life.

The Dance of the Gases

Our atmosphere isn't just a static mixture – it's a dynamic system in constant motion. The air around us is always moving, driven by differences in temperature, pressure, and Earth's rotation. This movement creates weather patterns, distributes moisture, and plays a crucial role in distributing both beneficial and harmful components of our air.

Temperature plays a particularly fascinating role in air quality. Warm air rises and cool air sinks, creating vertical mixing that usually helps disperse pollutants. However, sometimes a layer of warm air can settle above cooler air near the ground, creating what scientists call a temperature inversion. This acts like a lid on a pot, trapping pollutants close to the surface where we breathe.

These temperature inversions have played a role in some of history's worst air pollution disasters. The infamous London "peasoupers" of the early 20th century were made worse by temperature inversions trapping coal smoke near the ground. Similar conditions still cause air quality problems in cities worldwide, from Los Angeles to Beijing.

Geography's Role

The quality of the air we breathe is heavily influenced by the landscape around us. Mountains, valleys, large bodies of water, and even urban landscapes all affect how air moves and how pollutants disperse. Some locations are naturally prone to air quality problems because of their geography.

Consider Mexico City, built in a valley surrounded by mountains. This geographic setting, combined with the city's high altitude and intense sunlight, creates perfect conditions for forming ground-level ozone and trapping other pollutants. Los Angeles faces similar challenges, with mountains on three sides and the ocean on the fourth creating a natural bowl that can trap pollution.

Even the design of our cities affects air quality. Tall buildings can create "urban canyons" that trap pollutants at street level, while large areas of concrete and asphalt create "heat islands" that affect local air movement. Understanding these geographical effects is crucial for managing air quality and designing better cities.

The Human Impact

Humans have been affecting air quality since we first learned to use fire, but our impact has grown exponentially since the Industrial Revolution. The history of air pollution parallels the history of human technological development:

- **Ancient Rome:** Citizens complained about smoke from wood fires and metalworking
- **Medieval London:** The first laws restricting coal burning were passed

- **Industrial Revolution:** Cities became shrouded in coal smoke
- **Mid-20th Century:** Photochemical smog became a major problem in cities
- **Modern Era:** Complex mixtures of pollutants from various sources

The Great Smog of London in 1952 marked a turning point in our understanding of air pollution's health impacts. This disaster, which killed thousands in just a few days, led to some of the first modern air quality laws. It demonstrated that air pollution wasn't just an inconvenience – it could be deadly.

Your Personal Air Processing Plant

Your respiratory system is an incredibly sophisticated air processing plant that operates 24/7. With each breath, you take in about half a liter of air (more during exercise). This air travels down your windpipe and into increasingly smaller tubes in your lungs, eventually reaching tiny air sacs called alveoli.

These alveoli are where the magic happens – oxygen moves from the air into your blood, while carbon dioxide moves from your blood into the air to be breathed out. This gas exchange is crucial for every cell in your body. Your brain alone uses about 20% of your body's oxygen supply, despite making up only 2% of your body weight.

But your lungs don't just process oxygen and carbon dioxide. They have to deal with everything else in the air too, including:

- Particulate matter from various sources
- Volatile organic compounds
- Industrial emissions
- Vehicle exhaust
- Pollen and other allergens
- Microplastics
- Smoke from wildfires
- Chemical fumes from products

Your respiratory system has several defenses against airborne contaminants, including mucus that traps particles and tiny hair-like structures called cilia that help move these particles out. However, many modern pollutants can overwhelm or bypass these defenses, leading to various health problems.

The Global Air Connection

One of the most important things to understand about air quality is that air pollution knows no borders. The atmosphere is a single, interconnected system that circulates pollution around the globe. Dust from the Sahara Desert can affect air quality in the Caribbean. Smoke from wildfires in California can influence air quality in Europe. Industrial emissions from Asia can impact air quality in North America.

This global connection has important implications for air quality management:

- Local actions can have global consequences
- Solving air quality problems requires international cooperation
- Air quality standards need to be coordinated across borders
- Monitoring needs to occur on a global scale

Progress and Challenges

The good news is that we've made considerable progress in improving air quality in many parts of the world. In developed countries, air quality is generally better now than it was 50 years ago, thanks to:

- Environmental regulations
- Improved industrial technologies
- Cleaner vehicles
- Better understanding of air quality impacts
- More effective monitoring systems

However, significant challenges remain:

- Many developing countries are experiencing worsening air quality
- Climate change is affecting air quality in complex ways
- New sources of pollution are emerging
- Indoor air quality is often worse than outdoor air

- Population growth and urbanization create new challenges

Taking Action

Understanding air quality is the first step toward improving it. While the challenges can seem overwhelming, there are many actions we can take at both individual and collective levels:

Individual Actions:

- Monitor local air quality
- Use air purifiers in homes
- Reduce personal emissions
- Support clean air initiatives
- Choose clean transportation options

Collective Actions:

- Implement stronger air quality regulations
- Invest in clean energy
- Improve public transportation
- Protect and expand green spaces
- Support research and monitoring

The Future of Air

As we look to the future, several trends will shape the air we breathe:

- Increasing urbanization
- Climate change impacts
- Modern technologies for monitoring and control
- Changes in energy systems
- Evolving transportation systems
- Growing awareness of air quality impacts

The quality of our air will depend on the choices we make today. Will we commit to clean energy? Will we design cities that promote clean air? Will we take the steps needed to protect this invisible but essential resource?

Conclusion

The air we breathe is our most immediate connection to the environment around us. Every breath connects us to the atmosphere's ancient history and its modern challenges. Understanding what's in our air, how it affects us, and how we affect it is crucial for protecting this vital resource.

Remember that first breath of mountain air we talked about at the beginning? That's not just a pleasant experience – it's a reminder of what clean air feels like and what we should be striving for. Clean air isn't a luxury; it's a basic human right and a necessity for all life on Earth.

In the next chapter, we'll take a closer look at the specific pollutants that affect air quality and their impacts on human health and the environment. But for now, take a moment to appreciate the remarkable system that keeps you alive with every breath, and consider what you can do to help keep our shared air clean and healthy for all.

CHAPTER 2

NATURAL COMPONENTS AND CYCLES

Ever watched those nature documentaries where they show time-lapse footage of plants growing, clouds moving, and seasons changing? It's pretty mind-blowing to see months or years compressed into a few seconds. Well, if we could see the air around us in time-lapse, it would look like an incredibly complex dance of molecules, particles, and gases, all moving and changing in perfect cycles that have been running for billions of years.

Think of Earth's atmosphere as the world's biggest recycling system. Nothing is ever really "new" – it's all being constantly recycled, transformed, and reused. That oxygen molecule you just breathed in. It might have once been part of a dinosaur's breath, or maybe it was released by a tree in the Amazon rainforest last week. Pretty cool, right?

Let's dive into these natural cycles, but don't worry – I promise to keep this interesting and skip the boring chemistry lecture. Instead, imagine you're following the journey of these molecules as they travel through Earth's great recycling system.

First up: nitrogen. Remember from our last chapter that nitrogen makes up about 78% of our atmosphere. That's a lot of nitrogen! But here's the weird thing – most living things can't use nitrogen directly from the air. It's like having a pantry full of canned food but no can opener. The nitrogen needs to be "fixed" or converted into a form that plants and animals can use.

This is where the nitrogen cycle comes in, and it's pretty amazing. Lightning strikes (yes, really!) can fix nitrogen in the air, turning it into compounds that plants can use. But the real heroes are tiny bacteria that live in the soil and in the roots of certain plants like beans and peas. These microscopic chemists can take nitrogen from the air and turn it into plant food. It's like they've got tiny molecular can openers!

Here's how the nitrogen cycle works, in everyday terms:

- Those helpful bacteria convert nitrogen from the air into compounds in the soil
- Plants suck up these nitrogen compounds through their roots
- Animals eat the plants (or eat other animals that ate the plants)
- When plants and animals die or, ahem, produce waste, other bacteria break down the nitrogen compounds
- Some of these compounds go back into the air, and the cycle starts all over

Now, let's talk about everyone's favorite gas – oxygen! The oxygen cycle is like a perfectly choreographed dance between plants and animals. Plants take in carbon dioxide and water, use sunlight to make their food (that's photosynthesis – but you probably remember that from school), and release oxygen as a waste product. Lucky for us because we need that "waste" to survive!

We animals then breathe in that oxygen, use it to get energy from our food, and breathe out carbon dioxide – which the plants need. It's like we have a deal with plants: "You give us oxygen; we'll give you carbon dioxide." Talk about a perfect partnership!

But there's more to this cycle than just plants and animals. The oceans play a huge role too. Tiny marine plants called phytoplankton produce about half of Earth's oxygen. That's right – half of the oxygen you're breathing right now probably came from these microscopic ocean dwellers. Next time you're at the beach, take a moment to thank these tiny oxygen factories!

You might be wondering how these phytoplankton manage to produce so much oxygen. Well, they might be small, but there are trillions upon trillions of them in our oceans. They form massive blooms that can actually be seen from space! These blooms can cover hundreds of square miles of ocean surface. Think of them as underwater forests, working away day and night to keep our atmosphere stocked with oxygen.

But here's something that might surprise you – the oxygen we breathe today isn't the same oxygen that was around when Earth first formed. Back then, our atmosphere was completely different, with hardly any oxygen at all. It took billions of years of photosynthesizing organisms (starting with bacteria, then moving on to algae and eventually plants) to build up the oxygen-rich atmosphere we enjoy today. Talk about a long-term project!

Now, let's talk about something we can all relate to – water vapor and humidity. You know those days when your hair gets all frizzy and you feel like you're swimming through the air? That's water vapor in action! Water vapor is actually the third most abundant gas in our atmosphere, after nitrogen and oxygen, but its levels vary a lot depending on where you are and what the weather's like.

The water cycle (or hydrologic cycle, if you want to sound fancy) is happening all around us, all the time. The sun heats up water in oceans, lakes, and rivers, turning it into vapor. This vapor rises into the air – a process called evaporation. As it rises, it cools and condenses into tiny droplets, forming clouds. When these droplets get heavy enough, they fall as rain or snow, and the cycle begins again.

But here's something fascinating about the water cycle that most people don't realize – it's not just about liquid water turning into vapor and back again. There's also something called sublimation, where ice turns directly into vapor without becoming liquid first. Ever notice how a tray of ice cubes in your freezer seems to shrink over time, even though it's still below freezing? That's sublimation in action! This process is super important in places like Antarctica, where massive amounts of ice turn directly into vapor, affecting weather patterns across the entire Southern Hemisphere.

And speaking of weather patterns, let's talk about how water vapor creates some truly spectacular weather phenomena. You know those towering thunderclouds that look like giant cauliflowers in the sky? They're formed by rising columns of warm, moist air – essentially water vapor – that cool and condense as they climb higher into the atmosphere. These clouds, called cumulonimbus, can reach heights of up to 60,000 feet! That's higher than most commercial airplanes fly.

But it's not just about rain and clouds. Water vapor plays a crucial role in our climate. It's actually the most important greenhouse gas (yes, more than carbon dioxide!). Without water vapor trapping heat in our atmosphere, Earth would be a frozen wasteland. But like many things in nature, it's all about balance.

Here's something mind-bending to think about: the amount of water on Earth has remained pretty much constant for millions of years. The water in your morning coffee might have once been inside a woolly mammoth, or part of an ancient sea where the first fish evolved. Water molecules are constantly being recycled through the atmosphere, oceans, land, and living things. It's the ultimate example of recycling!

Speaking of things in the air, let's talk about natural particulates – all those tiny solid and liquid particles floating around us. Even the cleanest "fresh" air contains millions of these particles in every breath. Some are visible – like dust, pollen, or sea spray. Others are so small you need an electron microscope to see them.

Here's a sampling of what's naturally floating around in the air:

- Salt crystals from sea spray
- Dust from deserts and soil
- Pollen from plants
- Spores from fungi
- Volcanic ash
- Smoke from natural fires
- Bacteria and other microorganisms
- Bits of plant and animal matter

But wait – there's more! Let's dive deeper into some of these fascinating particles. Take sea spray, for example. Every time a wave crashes on a beach or a bubble burst on the ocean surface, tiny droplets of seawater are launched into the air. As the water evaporates, it leaves behind microscopic salt crystals. These aren't just any salt crystals – they're complex cocktails of different minerals that play crucial roles in cloud formation and global weather patterns.

And what about desert dust? The Sahara Desert alone pumps out an estimated 182 million tons of dust into the atmosphere each year! This dust doesn't just float around aimlessly – it travels across entire oceans, fertilizing distant ecosystems. Amazonian rainforests, for instance, depend on phosphorus-rich dust from the Sahara to thrive. It's like Earth has its own international postal service, delivering nutrients across the globe through the air!

These particles play important roles in our environment. Dust from the Sahara Desert, for example, fertilizes the Amazon rainforest – it's literally carrying nutrients across an ocean! Sea salt particles help form clouds. Pollen allows plants to reproduce. Even the bacteria floating in the air are part of important ecological processes.

Let's zoom in even further on these airborne microorganisms because they're absolutely fascinating. Scientists estimate that there are between 40 million and 1.6 billion bacterial cells in a cubic meter of air! Some of these bacteria are just along for the ride, but others have adapted specifically to life in the atmosphere. They've developed protective shells to shield them from UV radiation and can enter a dormant state when conditions aren't favorable.

But not all natural particles are beneficial. Volcanic eruptions can pump millions of tons of ash and gases into the atmosphere, affecting climate and air quality worldwide. Natural forest fires produce smoke that can be harmful to breathe. Pollen causes allergies for many people. Nature isn't always gentle!

Speaking of volcanic eruptions, let's talk about their massive impact on our atmosphere. When a volcano erupts, it's not just throwing out ash and lava. It's also releasing vast amounts of gases like sulfur dioxide, carbon dioxide, and water vapor. One large eruption can temporarily cool the entire planet by creating a haze that reflects sunlight back into space. The 1991 eruption of Mount Pinatubo in the Philippines, for example, caused global temperatures to drop by about 0.5°C (0.9°F) for over a year!

Here's something mind-bending to think about: every breath you take has particles that have traveled thousands of miles. That dust particle you just inhaled. It might have come from a desert in China. That salt crystal? Maybe from waves crashing on a beach in Australia. The air connects us all in ways we rarely think about.

And these traveling particles don't just move horizontally — they also move up and down through different layers of the atmosphere. Some particles can reach the stratosphere, more than 30 kilometers (18 miles) above Earth's surface! Up there, they can influence ozone levels, affect how much sunlight reaches Earth's surface, and even impact global wind patterns.

Let's zoom in on a particularly interesting type of natural particle: biological particles, or "bioaerosols" if you want to impress your friends. These include living things like bacteria and fungi, and biological materials like pollen, pet dander, and tiny bits of plants and insects. Some of these microorganisms can live their entire lives floating in the air!

And when we say, "live their entire lives," we mean it! Scientists have discovered bacteria reproducing in clouds, using the water droplets as tiny incubators. These airborne microbes even have their own food webs – some bacteria feed on organic compounds in the air, while others prey on smaller microorganisms. It's like there's a whole microscopic ecosystem floating above our heads!

These airborne microorganisms aren't just passive passengers – they're actively involved in atmospheric processes. Some bacteria and fungi actually help form ice crystals in clouds, leading to snow and rain. They're like nature's cloud seeders! Others break down airborne pollutants, helping clean the air naturally.

Let's talk more about these ice-nucleating bacteria because they're absolutely incredible. These microscopic organisms produce special proteins that help water freeze at temperatures way above the normal freezing point. Without these bacteria, many clouds wouldn't produce rain or snow nearly as efficiently. Scientists are even studying these bacteria to develop better artificial snow-making techniques for ski resorts!

Speaking of cleaning the air, nature has some amazing air purification systems. Forests act like giant air filters, trapping particles and absorbing pollutants. Rain washes particles out of the air. Even the electrical charges in the atmosphere help clean the air by causing particles to clump together and fall to the ground.

The way forests clean the air is particularly fascinating. Tree leaves have tiny hair-like structures and sticky surfaces that trap airborne particles. A mature tree can capture more than 500 grams of particles in a single year! But it's not just about trapping particles – trees also release compounds called phytoncides that can kill harmful bacteria and fungi in the air. This is one reason why a walk in the forest feels so refreshing!

But perhaps the most impressive natural air purification system is the ocean. Seas and oceans produce those salt particles we mentioned earlier, which help form clouds and affect weather patterns. The ocean surface also absorbs vast amounts of gases from the atmosphere, helping regulate their concentrations. It's like Earth's built-in air conditioning system!

Let's dive deeper into how oceans regulate our atmosphere. The surface of the ocean is constantly exchanging gases with the air above it. When there's more of a gas in the air than in the water, the ocean absorbs it. When there's more in the water, some gets released into the air. This process helps keep atmospheric gas levels stable. The oceans have absorbed about 30% of all the extra carbon dioxide humans have added to the atmosphere through burning fossil fuels!

All these natural components and cycles have been in balance for millions of years. But here's where things get interesting (and a bit concerning): human activities are now affecting these natural cycles in significant ways. We're adding new chemicals to the air, changing the concentration of natural components, and disrupting cycles that have been stable for millennia.

For example, we're adding more nitrogen compounds to the environment through fertilizers and burning fossil fuels. We're increasing carbon dioxide levels faster than plants can absorb it. We're putting more particles into the air through industrial processes and changing land use. It's like we're conducting a giant experiment with Earth's atmosphere, and we're not entirely sure how it will turn out.

The changes we're making to these cycles can have some surprising and far-reaching effects. For instance, the extra nitrogen we're adding to the environment doesn't just affect the air – it runs off into rivers and oceans, causing massive algae blooms that can create "dead zones" where fish can't survive. And the extra carbon dioxide we're putting into the atmosphere isn't just warming the planet – it's also making our oceans more acidic, which threatens marine life.

But don't get discouraged! Understanding these natural cycles is the first step in protecting them. When we know how these systems work, we can better understand how to prevent damage and help nature keep its balance. Plus, nature is remarkably resilient – given a chance, these natural cycles can often recover and rebalance themselves.

Scientists are learning new things about these natural cycles all the time. For example, recent research has discovered that certain types of marine bacteria can produce tiny capsules of DNA that float through the air and influence cloud formation. Who knew that genetic material could affect our weather? And researchers have found that plants don't just passively absorb what's in the air – they actively sense and respond to changes in air quality, adjusting their behavior to cope with different conditions.

As we move forward in this book, we'll explore how human activities affect these natural cycles and what we can do to help protect them. But for now, take a moment to appreciate the incredible system that's running all around us, invisible but essential, constantly recycling and renewing the air we breathe.

Remember that nitrogen molecule we talked about at the beginning? It's still out there, maybe in the air, maybe in a plant, maybe in you – continuing its endless cycle through Earth's great recycling system. Every breath connects us to these ancient cycles, making us part of a process that's been running since life began on Earth. Each breath you take isn't just pulling in oxygen – it's connecting you to a vast, intricate web of life that spans the entire planet.

Think about it: every time you exhale, you're adding to the great atmospheric soup that surrounds our planet. The carbon dioxide you just breathed out might soon be absorbed by a tree in your backyard, or it might travel on wind currents to help a plant grow on the other side of the world. You're not just living on Earth – you're an active participant in its natural cycles.

In the next chapter, we'll look at how human activities have changed these natural cycles and what that means for our air quality. But for now, take another breath and think about the amazing journey those molecules have taken to reach your lungs. Consider how each molecule in that breath might have been part of countless other living things over millions of years. When you really think about it, every breath is a little miracle of nature's recycling system.

CHAPTER 3

MAJOR AIR POLLUTANTS

Remember when you were a kid, and your parents told you not to pick up stuff from the ground because it might be dirty? Well, it turns out the air has its own version of "dirty stuff" that we need to watch out for. These aren't just minor annoyances – they're serious troublemakers that the Environmental Protection Agency (EPA) tracks like a strict parent watching problem children.

Let's talk about what I like to call **"The Bad Air Gang"** – six major pollutants that the EPA considers so important they've given them special status as "criteria pollutants." Think of them as the most wanted list of air pollution. These troublemakers have been causing problems for decades, and it's about time you knew exactly who they are and what they're up to.

First up, let's meet the particle posse – or as scientists call it, particulate matter (PM). Now, these aren't just specks of dust you can see floating in a sunbeam. We're talking about particles so small you could fit thousands of them on the period at the end of this sentence. They come in two main sizes: PM10 (the bigger bullies) and PM2.5 (the really sneaky small ones).

Here's a way to think about particle sizes that might blow your mind: if a human hair is about 70 micrometers wide (that's really tiny), PM10 particles are less than 10 micrometers (even tinier), and PM2.5 particles are less than 2.5 micrometers (seriously tiny). The scary part? The smaller they are, the deeper they can get into your lungs.

To put this in perspective, let's imagine you're shrinking down to the size of these particles (kind of like in that old movie "Honey, I Shrunk the Kids"). At this scale, a grain of beach sand would look like a massive boulder, and a human hair would appear as wide as a large tree trunk. PM10 particles would be like basketballs bouncing around, while PM2.5 particles would be more like marbles. Now imagine all of these zooming through your airways – pretty wild, right?

These particles come from all sorts of places:

- Construction sites kicking up dust
- Cars and trucks spewing exhaust
- Factories sending smoke into the sky
- Fires burning everything from wood to trash
- Even natural sources like volcanoes and desert dust

And here's something that might surprise you – your favorite scented candle or that cozy fireplace? Yep, they're particle producers too. In fact, a study found that burning one scented candle can produce as many particles as you'd get from a diesel truck idling for an hour! Don't worry, though – we're not saying you need to give up all your candles. It's good to be aware and maybe save them for special occasions.

But here's the really tricky part about particles – they're not quite simple specks of dirt. They're more like tiny pollution sponges, often carrying other harmful chemicals along for the ride. It's like they're giving free piggyback rides to other pollutants straight into your lungs.

Think of these particles as the delivery service of the pollution world. Just like how a delivery truck might carry packages of all shapes and sizes, these particles can carry various nasty chemicals. Some might be carrying heavy metals, others could be coated with cancer-causing compounds, and some might even be harboring bacteria or viruses. It's like a microscopic toxic carpooling service that nobody asked for!

Now, let's move on to everyone's least favorite summer companion:

ground-level ozone, the main ingredient in smog. And no, this isn't the same ozone that protects us from UV rays up in the stratosphere – this is its troublemaking cousin that hangs out at ground level where we breathe.

It's kind of ironic, really – up in the stratosphere, ozone is our friend, blocking harmful UV rays like a planetary sunscreen. But down here at ground level, it's like that friend who was cool in high school but now just causes problems at every party. Same molecule, totally different impact depending on where it hangs out!

Here's the weird thing about ground-level ozone: it's not directly released by cars or factories. Instead, it's created when other pollutants (mainly nitrogen oxides and volatile organic compounds) get together and react in sunlight. It's like a chemical party where nobody invited us, but we all have to deal with the hangover.

Let me break down this chemical party for you. Imagine nitrogen oxides (NOx) and volatile organic compounds (VOCs) are like two groups of teenagers who behave okay on their own. But when they meet up in the presence of sunlight (think of it as their party venue), they start interacting in ways that create this troublemaker called ozone. It's like mixing red and blue paint to get purple, except in this case, we're getting something we really don't want!

You know those hot, sunny, still days when the air looks kind of brownish-yellow and feels thick? That's probably smog, and it's packed with ozone. It's worst during summer afternoons when the sun is beating down and there's not much wind. That's why you'll often hear weather forecasters talk about "ozone action days" – days when conditions are perfect for this unwanted chemical cookout.

Speaking of ozone action days, here's a pro tip:

If you're planning outdoor exercise, check your local air quality index first. On high ozone days, it's better to work out in the morning or evening when levels are lower. It's like planning your beach day around the tide – timing matters!

Moving on to our next bad guy:

Carbon monoxide (CO). This one's particularly sneaky because you can't see it, smell it, or taste it. It's like the ninja of air pollutants. But instead of cool ninja moves, it pulls a really nasty trick – it's better than oxygen at binding with your blood cells, which means it can literally steal your breath away.

Let's geek out about this for a minute because it's fascinating in a terrifying sort of way. Your blood cells have these special proteins called hemoglobin that normally carry oxygen throughout your body. But carbon monoxide is like that person who always cuts in line – it pushes its way in and bonds with hemoglobin about 200 times more strongly than oxygen does. Once CO is in, oxygen can't get a spot, and your cells start struggling to get the oxygen they need.

Carbon monoxide mainly comes from vehicles and anything else that burns fuel incompletely. That's why you should never run a generator or grill inside your house or garage. CO poisoning sends thousands of people to the hospital every year, and sadly, it kills hundreds. It's a reminder that just because you can't see a pollutant doesn't mean it can't hurt you.

Here's a scary fact, even your car sitting in your driveway can be a carbon monoxide hazard if it's running. On a frosty winter morning, don't warm up your car in the garage, even with the door open. That CO can build up faster than you'd think and seep into your house. And while we're at it, let's talk about keyless cars – there have been tragic cases where people accidentally left their cars running in attached garages, leading to CO poisoning. Always double-check that your car is off!

Next up are the sulfur and nitrogen compounds – or as I like to call them, the stinky siblings. Sulfur dioxide (SO2) has that classic rotten egg smell, while nitrogen oxides (NOx) contribute to that sharp, burning sensation, you might feel in your nose on highly polluted days.

Fun fact (well, maybe not so fun): that rotten egg smell from sulfur dioxide is the same compound that gives certain foods their distinctive odors. Love garlic? Thank organic sulfur compounds! Can't stand the smell of overcooked Brussels sprouts? Blame sulfur again! In nature, these compounds play important roles. It's only when we start pumping out too much of them that they become a problem.

These pollutants are mainly produced when we burn fossil fuels like coal and oil. Power plants are major sources, along with industrial facilities and vehicles (especially diesel engines). But these compounds don't just hang around in the air – they can transform into other pollutants and even come back down as acid rain.

Speaking of diesel engines, ever notice how some trucks release visible black smoke when they accelerate? That's particulate matter rich in nitrogen oxides. Some folks modify their diesel trucks to produce more of this smoke (they call it "rolling coal") – but here's the thing, it's not just showing off, it's literally throwing money away as unburned fuel while polluting the air we all breathe. Not cool, right?

Yeah, you heard that right – acid rain. When sulfur dioxide and nitrogen oxides mix with water in the atmosphere, they create acids that fall back to Earth in rain or snow. It's like these pollutants are doing a horrible magic trick, turning normal rain into something that can damage buildings, kill trees, and make lakes so acidic that fish can't survive.

The acid rain story is actually a fitting example of how air pollution doesn't respect borders. Back in the 1980s, Canada was getting really upset because acid rain from U.S. factories was damaging their forests and lakes. It led to some tense diplomatic moments but eventually resulted in stronger pollution controls on both sides of the border. These days, acid rain is much less severe in North America, though it's still a major problem in some parts of the world.

Last but definitely not least in our lineup of EPA's most wanted: lead and other heavy metals. Now, you might be thinking, "Wait, didn't we get rid of lead in gasoline years ago?" Yes, we did (mostly – it's still used in some aviation fuel), and that was one of the biggest environmental success stories ever. Lead levels in the air dropped by 98% after we took it out of gasoline.

The lead story is fascinating because it shows both how much damage pollution can do and how we can fix things when we try. Before lead was banned from gasoline, it was everywhere – in the air, soil, and even people's blood. Scientists found that kids growing up in that era had lower IQs and more behavior problems because of lead exposure. When lead was finally removed from gas, childhood blood lead levels plummeted, and researchers even found that crime rates dropped as those kids grew up! Talk about an unexpected bonus from environmental protection!

But lead is still out there, along with other heavy metals like mercury, arsenic, and cadmium. They come from industrial processes, waste incineration, and some types of mining operations. These metals are particularly nasty because they don't break down – they just keep cycling through the environment, potentially ending up in our food and water as well as our air.

Here's a mind-bending fact about mercury:

A single gram of mercury from a broken thermometer can contaminate a 20-acre lake to the point where the fish are unsafe to eat. That's because bacteria in the lake convert the mercury into an even more toxic form that accumulates up the food chain. So that mercury in the air from a coal-fired power plant? It could end up concentrating in the fish on your dinner plate years later.

Let's take a deeper dive into why these pollutants are such sad news for our health. When you breathe in particulate matter, the larger PM10 particles usually get caught in your nose and upper airways – kind of like your body's built-in air filter. That's why you might cough or sneeze when you're in a dusty place. But those sneaky PM2.5 particles? They can get deep into your lungs and even cross into your bloodstream.

Think of your respiratory system like a tree growing upside down in your chest. The trunk is your trachea, the branches are your bronchi, and the leaves are tiny air sacs called alveoli where oxygen enters your blood. PM10 particles mostly get stuck in the trunk and bigger branches, but PM2.5 particles can travel all the way to those leaves. Some can even squeeze through the walls of the air sacs and enter your bloodstream, where they can travel to your heart, brain, and other organs.

Once these tiny particles get into your system, they can cause all sorts of problems:

- Irritation of your eyes, nose, and throat
- Coughing and difficulty breathing
- Reduced lung function
- Irregular heartbeat

- Aggravated asthma
- And in the long term, they might even contribute to heart disease and lung cancer

Recent research has found even more concerning effects. Scientists have discovered that these particles might be linked to diabetes, depression, and even dementia. They've even found tiny pollution particles in the placentas of pregnant women – talk about starting pollution exposure early!

Ozone is another troublemaker when it comes to health effects. Think of it as a super-aggressive cleaner – it reacts with and damages any biological tissue it touches. When you breathe it in, it's like giving your lungs a chemical burn.

That's why breathing elevated levels of ozone can:

- Make you cough and feel short of breath
- Cause your chest to feel tight or painful
- Trigger asthma attacks
- Inflame and damage your airways
- Make it harder to fight off respiratory infections

Want to know something wild about ozone? It can actually change the way plants grow. High ozone levels can damage plant leaves and reduce crop yields. Some scientists estimate that ozone pollution reduces global crop production by billions of dollars each year. So, it's not just our lungs that suffer – it's also affecting the food we grow!

Carbon monoxide's health effects are particularly scary because they can happen so quickly. Remember how it competes with oxygen in your blood? Well, when CO wins that competition, it means your body's cells aren't getting the oxygen they need.

The results can include:

- Headaches and dizziness
- Nausea and vomiting
- Confusion and drowsiness
- And in severe cases, loss of consciousness and death

Here's a wild fact about carbon monoxide poisoning:

It can cause symptoms that look like food poisoning or the flu. That's why it's super important to have CO detectors in your home – they're like smoke detectors for this invisible threat. And if several people in the same building suddenly develop similar symptoms? That's a big red flag for possible CO poisoning.

The sulfur and nitrogen compounds are like bullies that pick on your respiratory system.

They can:

- Irritate your airways
- Make it harder to breathe
- Trigger asthma attacks
- Make you more susceptible to respiratory infections

- And over the long term, they might contribute to the development of chronic respiratory diseases

Did you know that your nose is actually really good at detecting sulfur dioxide? Most people can smell it at concentrations well below harmful levels. It's like your body's built-in warning system saying, "Hey, something's not right here!" Unfortunately, we can't smell many other pollutants, which is why we need scientific monitoring to keep track of them.

As for lead and other heavy metals, they're the pollutants that keep on giving – and not in a safe way.

These metals can build up in your body over time, potentially causing:

- Damage to your nervous system
- Kidney problems
- Blood disorders
- Developmental issues in children
- Memory and behavior problems

Here's something scary about lead:

Scientists now believe there is no safe level of exposure, especially for kids. Even tiny amounts can affect brain development. And once lead gets into your body, it can hide out in your bones for decades, potentially causing problems years after the initial exposure. It's like your body's unwanted time capsule of pollution.

But here's something important to remember, the effects of air pollution aren't distributed equally. Some people are more vulnerable than others.

Children breathe more air per pound of body weight than adults, so they get a bigger dose of whatever's in the air

Elderly people often have underlying health conditions that make them more sensitive to air pollution

People with asthma or other respiratory diseases are more affected by poor air quality

Low-income communities often face higher levels of pollution because they're more likely to be located near industrial facilities or major highways

This inequality in air pollution exposure is a prime example of environmental justice issues. Studies have shown that communities of color and low-income neighborhoods often face much higher pollution levels than wealthier, predominantly white areas. For instance, one study found that Black Americans are exposed to about 1.5 times more particulate matter than white Americans, regardless of income level or where they live.

The good news is that we've made progress in fighting these pollutants. Since the Clean Air Act was passed in 1970, levels of all six criteria pollutants have dropped significantly. Lead pollution is down 98%, carbon monoxide is down 77%, and even particulate matter has decreased by about 40%.

These improvements didn't happen by accident. They're the result of better technology, stricter regulations, and increased awareness. For example, catalytic converters in cars now remove most of the carbon monoxide and nitrogen oxides from vehicle exhaust. Power plants have installed scrubbers that catch sulfur dioxide before it can escape into the air. And newer diesel engines produce far fewer particles than older models.

The unwelcome news is that we still have work to do. Millions of Americans still live in areas that don't meet the EPA's air quality standards. And globally, air pollution is getting worse in many developing countries as they industrialize.

This global aspect of air pollution is crucial to understand. Air pollution doesn't respect national borders – particles from a dust storm in the Gobi Desert can travel across the Pacific Ocean and affect air quality in California. Smoke from wildfires in Canada can darken skies in New York City. And greenhouse gases emitted in one country contribute to climate change everywhere. That's why addressing air pollution requires international cooperation, not just local action.

Speaking of international impacts, let's talk about something called the "brown cloud" phenomenon. In parts of Asia, there's a massive layer of air pollution that can be seen from space! It's a mix of all the pollutants we've talked about, and it can be so thick it reduces sunlight reaching the ground by up to 15%. Imagine that – pollution so bad it's actually dimming the sun! This isn't just Asia's problem either – these pollution clouds can travel across oceans and affect air quality thousands of miles away.

But knowledge is power, right? Now that you know who these pollutants are and what they can do, you're better equipped to protect yourself and your family. And there's more good news – there are actually lots of things you can do to reduce your exposure to these bad actors:

- Check daily air quality forecasts just like you check the weather
- Plan outdoor activities for times when pollution levels are lower
- Use air purifiers in your home (and yes, some houseplants can help too!)
- Maintain your car's emission control systems
- Avoid exercising near high-traffic areas
- Keep your windows closed on high pollution days

Let's talk more about those air quality forecasts because they're super helpful but often misunderstood. The Air Quality Index (AQI) is like a pollution thermometer that goes from 0 to 500. But unlike a regular thermometer, lower numbers are better! A score below 50 is great, while anything above 150 is unhealthy for everyone. The index is color-coded too – green for good, yellow for moderate, orange for unhealthy for sensitive groups, and red for unhealthy. Purple and maroon? Those mean the air is really bad – time to stay inside!

And here's a cool tech tip, there are plenty of smartphone apps that can give you real-time air quality data for your location. Some even send alerts when pollution levels get high. It's like having a personal pollution watchdog in your pocket! Just be sure to get your data from reliable sources like AirNow.gov or your local air quality management district.

In our next chapter, we'll look at some even sneakier pollutants – the ones that might be lurking inside your home right now. You might be surprised to learn that indoor air can actually be more polluted than outdoor air! Think about it – we spend about 90% of our time indoors, so even low levels of indoor pollutants can add up to significant exposure over time.

But before we move on, here's something to think about:

Every breath you take is about 0.5 liters of air. The average person takes about 20,000 breaths per day. That means you're processing about 10,000 liters of air through your lungs every day! With numbers like that, it's easy to see why air quality is so important for our health.

And here's one final thought that might blow your mind the air you're breathing right now is connected to every other breath being taken on Earth. The atmosphere doesn't care about national borders or social boundaries – it's one big, shared resource that we all depend on. When we improve air quality, we're not just helping ourselves – we're helping everyone.

Remember those six criteria pollutants we started with? They're like the beginning when it comes to air pollution. Scientists are constantly discovering new air pollutants and learning more about how they affect our health. But don't let that discourage you! Every step we take to reduce pollution helps, whether it's walking instead of driving, choosing clean energy, or supporting policies that protect air quality.

So, take a deep breath (preferably of clean air) and give yourself a pat on the back for becoming more air-quality aware. Knowledge really is the first step toward cleaner air for everyone. And next time someone asks you about air pollution, you can impress them with your understanding of The Bad Air Gang and all their troublemaking ways!

Just remember – while we've covered some serious topics here, there's always hope. The improvement in air quality since the 1970s shows that we can make positive changes when we work together. Every day, scientists are developing latest technologies to reduce pollution, and more people are becoming aware of the importance of clean air. You're now part of that informed community, ready to make a difference in the quality of the air we all share.

Stay tuned for our next chapter, where we'll explore the hidden world of indoor air pollution. Trust me – you'll never look at your vacuum cleaner the same way again!

CHAPTER 4

CHEMICAL THREATS IN OUR AIR

Remember that "new car smell" everyone seems to love? Or how about the fresh scent of just-cleaned carpets? Well, I hate to be the bearer of bad news, but those familiar smells are actually telling us something's not quite right with our air. They're signs that we're breathing in volatile organic compounds, or VOCs – and trust me, there's nothing "organic" about them in the healthy, farmers-market sense of the word.

Think of VOCs as the gossips of the chemical world – they just can't keep to themselves. These chemicals love to spread around, turning from liquids or solids into gases at room temperature. That's what "volatile" means – they're eager to get into the air and crash your breathing party.

Let's dig deeper into that new car smell for a moment, because it's a perfect example of how complex chemical exposure can be. That distinctive scent actually comes from dozens of different chemicals slowly releasing from the car's interior materials. We're talking about plasticizers from the dashboard, flame retardants from the seats, adhesives from the carpeting, and various other compounds from all the synthetic materials used in modern vehicles. Some studies have found over 200 different chemicals contributing to that sought-after scent!

You might be surprised to learn just how many everyday items are releasing VOCs into your air:

- That fresh paint on your walls
- Your new memory foam mattress
- The dry-cleaned clothes in your closet
- Your shower curtain (especially if it's vinyl)
- That air freshener you just sprayed
- The cleaning products under your sink
- Even that printer churning out documents at work

And here's something that might blow your mind – even your morning cup of coffee releases VOCs! That wonderful coffee aroma? It's actually a complex mixture of volatile organic compounds. The difference is these natural VOCs aren't typically harmful like their synthetic cousins. It's a good reminder that not all chemicals are bad – it's about which ones, how much, and in what context.

One of the most common VOCs is formaldehyde. Yes, the same stuff used in high school biology to preserve frogs for dissection. It's found in pressed wood products, some types of insulation, and many household products. Other common VOCs include benzene (from gasoline and tobacco smoke), perchloroethylene (from dry cleaning), and methylene chloride (from paint strippers).

Let's take a closer look at formaldehyde because it's such a sneaky chemical. It's used in so many products because it's an excellent preservative and bonding agent. That particle board furniture from your favorite Swedish retailer? The glue holding it together probably contains formaldehyde. Your wrinkle-free shirts? They likely got that way through a formaldehyde treatment. Even some cosmetics contain formaldehyde or formaldehyde-releasing preservatives. It's like this chemical has managed to infiltrate every corner of our lives!

The tricky thing about VOCs is that they can cause both immediate and long-term health effects. Short-term exposure might give you:

- Headaches
- Dizziness
- Eye and throat irritation
- Nausea
- That generally icky feeling like you're coming down with something

And here's something particularly concerning: many people experience these symptoms without realizing that chemicals in their air are the cause. They might blame it on stress, lack of sleep, or coming down with a cold. It's like having a toxic roommate who's gaslighting you about why you're feeling bad!

Long-term exposure? That's when things get really concerning:

- Liver damage
- Kidney problems
- Central nervous system issues
- And some VOCs are known or suspected to cause cancer

What makes this even scarier is that the effects of long-term exposure often don't show up until years later. It's like those VOCs are playing the long game, slowly but steadily affecting your health over time.

Now let's shift gears and talk about industrial emissions – the big leagues of air pollution. If VOCs are like neighborhood gossips, industrial emissions are like bullhorns broadcasting chemicals everywhere. These come from factories, power plants, refineries, and other industrial facilities.

Here's a mind-boggling fact:

According to the EPA's Toxic Release Inventory, industrial facilities in the United States release billions of pounds of chemicals into the air each year. And those are just the releases that are reported! Many smaller facilities and certain types of emissions don't have to be reported at all. It's like having a leaky faucet in your house, but not being able to measure all the water that's dripping out.

The scale of industrial emissions can be mind-boggling. A single large factory might release tons of chemicals into the air every year. These can include:

- Sulfur dioxide from burning coal and oil
- Nitrogen oxides from manufacturing processes
- Heavy metals from smelting operations
- A whole alphabet soup of organic chemicals from various industrial processes

And here's something that might surprise you – some of these industrial chemicals can actually create their own weather! In areas with heavy industrial activity, the combination of particles and chemicals in the air can increase fog formation and even affect local rainfall patterns. It's like these facilities are accidentally doing their own form of weather modification.

But here's something that might surprise you – industrial emissions aren't just coming from those obvious smoking chimneys. Lots of chemicals escape during regular operations through what's called "fugitive emissions." These are the leaks, spills, and evaporation that happen during normal industrial processes. It's like having a leaky faucet, except instead of wasting water, it's releasing chemicals into the air.

To understand fugitive emissions better, think about a game of musical chairs where the chairs are molecules of chemicals, and the players are the containers and pipes meant to hold them. No matter how well you plan, some players (chemicals) are always going to end up without a chair (container). In industrial settings, these escapees can add up to significant amounts of pollution that never goes through any kind of treatment or filtering system.

Let's move to the farm fields now. You might think agricultural air pollution is just about dust from plowing, but there's a lot more going on out there.

Modern farming often involves a complex cocktail of chemicals:

- Pesticides that drift through the air
- Ammonia from fertilizers and animal waste
- Methane from livestock and decomposing organic matter
- Various chemicals from burning agricultural waste

Did you know that livestock farming is one of the largest sources of methane emissions globally? A single dairy cow can produce about 220 pounds of methane per year through its digestive processes. Multiply that by the millions of cattle worldwide, and you've got a significant contribution to greenhouse gases. Who knew cow burps could be such a big deal for air quality?

Pesticide drift is particularly concerning. When farmers spray their crops, not all of the chemicals stay where they're supposed to. Some of it can travel for miles on the wind, affecting communities far from the original application site.

It's like having a neighbor water their lawn on a windy day –
except instead of getting splashed with water, you're getting
doused with chemicals.

Here's a scary thought about pesticide drift:

Scientists have found these chemicals in rain samples
collected in areas far from any farming activity. Some studies
have even detected agricultural pesticides in Arctic ice cores!
This shows how these chemicals can travel through the
atmosphere and end up literally anywhere on Earth. It's a
powerful reminder that when it comes to air pollution, what
goes up must come down – somewhere.

Speaking of chemicals traveling where they shouldn't, let's
talk about vehicle exhaust. Every day, millions of cars, trucks,
buses, and other vehicles are pumping out a complex mixture
of chemicals into our air. It's not just carbon monoxide and
carbon dioxide – vehicle exhaust contains:

- Nitrogen oxides
- Particulate matter (especially from diesel engines)
- Volatile organic compounds
- Benzene and other aromatic compounds
- Formaldehyde and other aldehydes

Want to hear something wild about vehicle exhaust?
Scientists have found that the air inside your car can actually
be more polluted than the air outside, especially when you're
stuck in traffic. Those exhaust fumes from the vehicles

around you can get sucked into your car's ventilation system. It's like sitting in a box collecting all the neighborhood's pollution!

And here's where things get really interesting (or scary, depending on how you look at it): all these chemicals don't just hang out in the air by themselves. They interact with each other and with sunlight in ways that can create entirely new pollutants. It's like having a chemistry lab running 24/7 right above our heads.

These chemical interactions can be incredibly complex. Scientists have identified thousands of different chemical reactions happening in urban air. Some pollutants break down into less harmful compounds (thank goodness!), but others can combine to form more dangerous ones. It's like a chemical version of musical chairs, but with much higher stakes.

One of the most important of these atmospheric interactions is the formation of smog. When nitrogen oxides and VOCs from vehicles and industry mix with sunlight, they create ground-level ozone and other nasty compounds. It's a perfect example of how pollutants can team up to become even more troublesome.

Here's something fascinating about smog formation: it often peaks in the afternoon, not during rush hour when emissions are highest. That's because the chemical reactions need sunlight to happen. So, while the morning commute might pump out lots of pollutants, it's the afternoon sun that transforms them into smog. Nature's timing isn't always on our side!

But the chemical interactions don't stop there. Some pollutants can attach themselves to particles in the air, hitching a ride deep into our lungs. Others can react with water vapor to form acid rain. Some can even break down into different chemicals that might be more harmful than the original pollutants.

Think of it like a chemical game of tag, where pollutants are constantly combining, breaking apart, and transforming. Except in this game, we're all unwitting players, breathing in whatever combinations end up in our air. And just like in any game, some combinations are worse than others.

Here's a disturbing thought: scientists are still discovering new chemical reactions happening in our atmosphere. It's like we've been running a giant, uncontrolled experiment with our air, and we're still learning about the results.

For example, researchers recently discovered that some common cleaning products can react with skin oils left on indoor surfaces to create new types of indoor air pollutants. Who would have thought that cleaning your house could actually create new forms of pollution? It's like trying to sweep up dust only to have it transform into something else entirely.

Take, for example, the way some chemicals can hang around in the environment for years or even decades. These "persistent" pollutants can travel around the globe on air currents, ending up in places far from where they were released. Scientists have found industrial chemicals in Arctic ice and remote mountain lakes – places that have never seen a factory or a car.

These persistent pollutants are like the plastic of the chemical world – they just don't break down easily. Some can remain in the environment for decades or even centuries. DDT, for instance, was banned in the United States in 1972, but it's still showing up in air samples today. Talk about overstaying your welcome!

But before you start thinking about wearing a gas mask 24/7, let's talk about what we can actually do about all this.

First, awareness is key. Now that you know about these chemical threats, you can make better choices about:

- The products you use in your home
- When and where you spend time outdoors
- How you maintain your vehicle
- What kinds of cleaning products you buy
- Even how you store chemicals in your garage

And here's some good news: there are more eco-friendly alternatives available now than ever before. From natural cleaning products to low-VOC paints, manufacturers are responding to consumer demand for safer options. It's like voting with your wallet for cleaner air!

Many chemical exposures are higher indoors than outdoors, which means we have more control over them than you might think.

Simple actions like:

- Opening windows for ventilation
- Choosing low-VOC products
- Avoiding air fresheners
- Not idling your car in the garage
- Reading product labels carefully

These can all help reduce your exposure to harmful chemicals.

Let's talk about ventilation for a moment because it's super important but often overlooked. Most people don't realize that indoor air is typically 2-5 times more polluted than outdoor air, and sometimes up to 100 times worse! That's because chemicals released indoors get concentrated in our sealed-up buildings. Opening windows for just 10 minutes a day can make an enormous difference.

On a larger scale, we've made progress in reducing some chemical threats to our air. Environmental regulations have forced industries to clean up their acts somewhat. Many dangerous chemicals have been banned or restricted. Vehicle emission standards have gotten stricter.

Remember leaded gasoline? That's a remarkable success story. When we finally got lead out of gasoline, blood lead levels in children dropped dramatically. It's proof that when we recognize a chemical threat and take action, we can make real improvements in air quality and public health.

But there's still work to do. Thousands of new chemicals are created every year, and many of them make their way into the air before we fully understand their effects. It's like we're playing catch-up in a game where the rules keep changing.

Here's a sobering statistic: of the more than 80,000 chemicals registered for use in the United States, only a small fraction has been thoroughly tested for their health effects. It's like we're conducting a massive experiment with our air and our health, and we're all involuntary participants.

The good news is that technology for detecting and controlling air pollution keeps improving. We have better monitoring systems, more efficient filters, and cleaner industrial processes. We're getting better at understanding how chemicals interact in the atmosphere and what that means for our health.

Modern air quality monitors can now detect pollutants at incredibly low levels – parts per billion or even parts per trillion. That's like being able to find a single drop of water in an Olympic-sized swimming pool! This improved detection helps us better understand what's in our air and how to clean it up.

Remember that "new car smell" we talked about at the beginning? Many car manufacturers are now working to reduce the VOCs in their vehicles. Some furniture makers are using fewer toxic materials. Even paint companies are developing low-VOC alternatives.

And here's something exciting: some companies are developing "smart" materials that can actually help clean the air. There are now paints and fabrics that can break down air pollutants, and buildings designed to filter and clean the air naturally. It's like turning our built environment into a giant air purifier!

But perhaps the most important thing to remember is this: while chemical threats in our air are serious, they're not unstoppable. Every time we choose a less toxic product, support cleaner technologies, or speak up for better air quality regulations, we're helping to clear the air.

Think of it like voting – every small action counts. When you choose low-VOC paint for your bedroom, you're not just protecting your own air quality; you're also sending a message to manufacturers that consumers want safer products. When you maintain your car properly, you're not just reducing your own emissions; you're contributing to cleaner air for everyone.

In the next chapter, we'll look at how these chemical threats specifically affect our health and what we can do to protect ourselves. But for now, take a moment to think about the choices you make that might affect air quality. Because when it comes to chemical threats in our air, we're all part of both the problem and the solution. And remember – every breath of cleaner air is a step in the right direction!

Let's talk about some other chemical threats that might be lurking in our air. Ever heard of PFAS? They're called "forever chemicals" because they basically never break down in the environment. These perfluoroalkyl and polyfluoroalkyl substances are used in everything from non-stick cookware to water-resistant clothing, and yes, they can get into our air.

Think about your last camping trip when you waterproofed your tent or boots. That spray you used. It likely contained PFAS. When you use these products, these chemicals don't just stay on the surface – they get into the air, and from there, they can get into your lungs. And remember that "forever" part? Once they're in your body, they tend to stay there. Scientists have found PFAS in the blood of people all over the world, including in remote Arctic communities!

And speaking of synthetic materials, let's talk about microfibers. When you think of air pollution, you probably don't think about your clothes, right? Well, it turns out that synthetic fabrics like polyester and nylon release tiny fibers into the air every time we wear, wash, or dry them. These microfibers are so small they can float in the air and yes, we can breathe them in. One study found that we might be breathing in up to 68,000 microfibers per year just from our clothes and textiles!

Here's another troublemaker you might not have heard about phthalates. These chemicals are used to make plastics soft and flexible, and they're everywhere – in vinyl flooring, shower curtains, plastic packaging, and even children's toys. The problem? They don't stay put. Phthalates gradually release into the air, especially when these materials get warm. Ever noticed how your car gets super hot inside on a sunny day? That heat is basically cooking phthalates out of your car's interior materials and into the air you breathe.

Let's talk about something that hits close to home – literally. Flame retardants are chemicals added to furniture, electronics, and building materials to prevent fires. Sounds good, right? The problem is these chemicals don't stay in the products. They slowly release into the air and attach to dust particles. When you plop down on your couch or office chair, you're probably releasing a tiny puff of flame retardant-laden dust into the air. And guess what? Some of these chemicals have been linked to cancer, hormone disruption, and developmental problems.

And here's a seasonal threat many people don't think about – mothballs. They might keep moths from eating your winter sweaters, but that distinctive smell is actually the scent of naphthalene or paradichlorobenzene slowly turning into gas and escaping into your air. Both these chemicals are suspected carcinogens. It's like trading clothes-eating moths for an invisible chemical threat!

Ever think about what happens when you burn plastics? Whether it's accidental or in waste incinerators, burning plastic releases, some seriously nasty chemicals called dioxins. These are some of the most toxic chemicals known to science, and once they're in the environment, they stick around for a long time and can travel long distances through the air.

Even low-level exposure to dioxins can cause reproductive and developmental problems.

Here's one that might surprise you – radon. It's a naturally occurring radioactive gas that can seep into buildings from the ground. You can't see, smell, or taste it, but it's the second leading cause of lung cancer in the United States after smoking. The scary part? It could be building up in your basement right now and you'd have no idea without testing for it.

Let's talk about another invisible threat – mercury vapor. We're not just talking about broken thermometers here. Mercury is released into the air from coal-burning power plants, certain types of manufacturing, and even crematoriums (because of mercury in dental fillings). Once it's in the air, it can travel long distances before falling back to Earth in rain or snow. Then it can get into the food chain, especially through fish, and end up back in our bodies.

And here's a modern threat that's becoming more common – 3D printer emissions. As 3D printing becomes more popular in homes and offices, researchers are finding that these machines release a complex mixture of VOCs and ultrafine particles during printing. It's like having a tiny factory running on your desk! Some of the chemicals released are known to be harmful, especially when printing with ABS plastic.

Let's not forget about quaternary ammonium compounds, or "quats" for short. These chemicals are in many disinfectant products, and their use skyrocketed during the COVID-19 pandemic. While they're great at killing germs, they can irritate your lungs and may contribute to respiratory problems like asthma. It's a classic case of the solution becoming part of the problem.

Here's a seasonal threat that affects millions: smoke from wildfires. As climate change makes wildfires more frequent and intense, more people are being exposed to smoke that can contain hundreds of chemical compounds. And it's not just the smoke you can see – wildfire smoke can react with other pollutants in the air to create new chemical threats.

When it comes to protecting yourself from these chemical threats, knowledge really is power.

Here are some added steps you can take:

- Test your home for radon – it's easy and inexpensive

- Choose natural fiber clothing, when possible, to reduce microfiber release
- Skip the mothballs and use cedar or lavender instead
- Avoid "antibacterial" products with quats when regular soap will do
- Keep your 3D printer in a well-ventilated area
- Use a HEPA filter with activated carbon to remove both particles and gases

And remember – reducing your exposure to chemical threats isn't about living in fear or becoming obsessed with every possible danger. It's about making informed choices when you can and taking reasonable precautions to protect yourself and your family.

CHAPTER 5

THE INDOOR AIR CRISIS

Ah, home sweet home. Your safe haven from the world. The place where you kick off your shoes, relax on the couch, and take a deep, comfortable breath... of potentially polluted air. Plot twist! That couch you're sitting on. It might be releasing flame retardants into the air. That deep breath? You might be inhaling more pollutants than you would outside on the sidewalk.

Here's a shocking truth: indoor air is typically two to five times more polluted than outdoor air, and in some cases, it can be up to 100 times worse. Pretty scary when you consider that most of us spend about 90% of our time indoors. That's right – we're hanging out in our most polluted environments for most of our lives.

Let's put those numbers in perspective for a moment. If you're following the typical American lifestyle, you're spending about 21.6 hours per day indoors. That's about 7,884 hours per year breathing indoor air! For comparison, you probably only spend about 2 hours a day outdoors. So even if indoor air was just slightly more polluted than outdoor air (which it usually isn't), your total exposure to indoor pollutants would still be much higher simply because of how much time you spend inside.

Think of your home as a terrarium. Just like those little glass gardens, your house traps and concentrates whatever's in the air. But unlike a terrarium, your home is full of synthetic materials, cleaning products, and other modern conveniences that can make the air quality worse than you'd find in any natural environment.

And just like a terrarium, your home has its own microclimate. The temperature, humidity, and air circulation patterns in your house can all affect how pollutants build up and move around. Warm air rises, carrying pollutants to upper floors. Cold windows create downdrafts that can stir up dust. Even your HVAC system can create "dead zones" where air doesn't circulate well, allowing pollutants to concentrate.

Let's start our indoor air quality tour in a typical living room. Look around yours right now. What do you see?

That comfy couch? Probably contains flame retardants that can get into the air. And we're not talking about just one type of flame retardant – furniture manufacturers often use a cocktail of different chemicals, each with its own potential health effects. These chemicals don't stay put either; they attach to dust particles and get into the air whenever you sit down or fluff your cushions.

The carpet? Likely off-gassing VOCs from its backing and treatments.

Modern carpets can contain dozens of different chemicals, including:

- Styrene from the backing material
- Formaldehyde from adhesives
- 4-phenylcyclohexene (that "new carpet smell")
- Stain-resistant treatments containing PFAS
- Antimicrobial treatments
- Moth-proofing chemicals

Those vinyl blinds? Releasing phthalates (chemicals that make plastic flexible). Phthalates are particularly concerning because they're endocrine disruptors, meaning they can interfere with your hormone system. And they don't stay in the blinds forever – they gradually migrate out into the air and dust, especially when the blinds get warm from sunlight.

That flat-screen TV? Getting warm and releasing flame retardants and other chemicals. Electronic devices don't just release chemicals when they're new – they can continue to emit pollutants throughout their lifetime, especially when they heat up during use. This is called thermally induced off-gassing.

That scented candle. Adding particulate matter and artificial fragrances to your air. Did you know that scented candles can release as many as 100 different chemicals into your air?

The fragrances themselves are usually proprietary blends, which means manufacturers don't have to disclose what's in them. And when you burn candles, you're not just releasing fragrance – you're also creating soot particles that can get deep into your lungs.

Move into the kitchen, and things get even more interesting:

Gas stove? Hello, nitrogen dioxide and carbon monoxide. Recent studies have found that gas stoves can release concerning levels of pollutants even when they're turned off! The blue flame you see when cooking might look clean, but it's actually creating a variety of pollutants through the combustion process:

- Nitrogen dioxide (NO2) that can irritate your airways
- Carbon monoxide (CO) that can cause headaches and dizziness
- Formaldehyde from incomplete combustion
- Ultrafine particles that can get deep into your lungs
- Methane leaks that contribute to climate change

Nonstick pans? They can release toxic fumes when overheated. Those convenient nonstick coatings start breaking down at temperatures above 500°F (260°C), releasing chemicals that can cause flu-like symptoms known as "polymer fume fever." And if you have a bird as a pet, these fumes can be lethal to them – that's why canaries were used in coal mines to detect dangerous gases.

Air freshener plug-in? Pumping artificial chemicals into your air 24/7. These devices typically use heat to release fragrances, which can react with ozone in your air to form formaldehyde and ultrafine particles. Some studies have found that using air fresheners can expose you to the equivalent of secondhand smoke in terms of certain pollutants!

That lemon-fresh cleaner? Probably contains harsh chemicals that linger in the air. That fresh citrus smell might make you think it's natural, but most commercial cleaners use artificial fragrances that can contain dozens of undisclosed chemicals. Plus, many cleaning products contain:

- Quaternary ammonium compounds ("quats") that can trigger asthma
- Chlorine-based disinfectants that release fumes
- Glycol ethers that can cause reproductive problems
- Ammonia that irritates eyes and lungs
- Pine or citrus oils that can react with ozone to form formaldehyde

Self-cleaning oven? Creates some seriously nasty fumes during the cleaning cycle. The elevated temperatures (around 900°F/482°C) used in the self-cleaning process can:

- Break down the nonstick coating inside the oven
- Create carbon monoxide from burning food residue

- Release fumes from cleaning products trapped in the oven
- Generate fine particles that can trigger asthma
- Create such elevated levels of pollutants that some manufacturers recommend leaving your house during the cleaning cycle!

The bathroom's no better:

Hair spray? VOCs galore. Aerosol products are particularly problematic because they create a fine mist that's easy to inhale. Hair sprays often contain:

- Propellants like propane or butane
- Vinyl acetate or other binding polymers
- Artificial fragrances
- Phthalates
- Preservatives like parabens and those particles can hang in the air for hours after you spray!

Air freshener? More artificial chemicals. Bathroom air fresheners are often stronger than those used in other rooms because manufacturers assume you're trying to cover up stronger odors.

This means you're getting exposed to higher levels of:

- Synthetic musks
- Phthalates used to make fragrances last longer
- Volatile organic compounds

- Allergens
- Masking agents that numb your sense of smell

Bleach-based cleaners? Releasing chlorine compounds. When you clean with bleach, you're not just killing germs – **you're also creating:**

- Chlorine gas that irritates your lungs
- Chloroform when bleach reacts with organic matter
- Chloramines if bleach mixes with ammonia or urine
- Hypochlorous acid that can trigger asthma
- Various chlorinated organic compounds

Hot shower steam? Helping release chemicals from all your bath products. Steam doesn't just make your bathroom mirror foggy – it also:

- Increases the rate at which chemicals are released from products
- Creates tiny water droplets that can carry chemicals deep into your lungs
- Makes your skin more permeable to chemicals
- Helps create chloroform from chlorinated water
- Promotes mold growth if ventilation is poor

Mold behind the tiles? Sending spores into the air. Bathroom mold isn't just unsightly – it can release:

- Allergenic spores
- Mycotoxins that can cause respiratory problems

- Musty odors that indicate microbial volatile organic compounds (MVOCs)
- Beta glucans that can trigger inflammation
- Spores that can colonize other areas of your home

And don't even get me started on the garage (if it's attached to your house):

Stored paints and solvents? Off-gassing even with the lids on. Paint cans and other containers aren't completely airtight, **so chemicals can escape through:**

- Vapor permeation through plastic containers
- Microscopic gaps around lids
- Temperature-induced pressure changes
- Damaged or corroded containers
- Spills and residues on container surfaces

Car exhaust? Seeping into your home every time you start the engine. Your garage can funnel car exhaust directly into your living space through:

- Gaps around doors
- Shared ventilation systems
- Air pressure differences
- Cracks in walls
- Utility penetrations

Lawn care chemicals? Releasing fumes as they sit on the shelf. Those garden products aren't just dangerous when you're using them – they're constantly releasing lesser amounts of:

- Pesticide vapors
- Herbicide fumes
- Fertilizer dust
- Fungicide residues
- Solvent vapors from liquid products

That bottle of antifreeze? Adding its own special blend to the air. Automotive products often contain highly volatile **chemicals that can evaporate even at room temperature, including:**

- Ethylene glycol from antifreeze
- Petroleum distillates from oils and lubricants
- Benzene and other aromatic compounds
- Methanol from windshield washer fluid
- Various cleaning solvents

But here's where things get really interesting: let's talk about building materials. Your house itself could be a source of indoor air pollution. Modern buildings are full of materials that can release chemicals into the air, and what's worse, these materials are literally all around you – in your walls, under your feet, and over your head.

Let's break down these building materials in detail:

Pressed wood products (like cabinets, furniture, and subflooring):

- Release formaldehyde from urea-formaldehyde resins used as adhesives
- Can off-gas for years, especially when new
- Emission rates increase with:
- Temperature (each 10°F increase doubles emissions)
- Humidity (higher moisture levels accelerate release)
- Age (newer materials emit more)
- Surface damage (scratches and cuts expose more material)

Different types have varying emission levels:

- Particleboard (highest emissions)
- Medium-density fiberboard (MDF)
- Hardwood plywood
- Oriented strand board (OSB)

Insulation

It's not just about keeping warm or cool:

Fiberglass particles can get into the air through:

- Gaps in walls and ceilings
- During installation or renovation
- Through air ducts
- When disturbed by vibration

- Some types contain formaldehyde-based binders

Can harbor moisture leading to:

- Mold growth
- Bacterial colonization
- Degradation of materials
- Release of chemical binders

Different types have different issues:

- Spray foam can off-gas for weeks after installation
- Cellulose may contain fire retardants
- Rock wool can release mineral fibers
- Foam board can release flame retardants

Paint

It's not just about color anymore:

Traditional paints release VOCs including:

- Benzene
- Toluene
- Xylene
- Ethyl acetate
- Glycol ethers

Can continue off-gassing long after they're dry through:

- Primary emissions (during drying)
- Secondary emissions (chemical reactions over time)
- Temperature-induced releases

- UV-induced breakdown

Even "zero-VOC" paints aren't completely chemical-free because:

- Colorants can add VOCs
- Biocides are often present
- Preservatives are needed
- Binding agents may release compounds

Different types have different impacts:

- Oil-based (highest emissions)
- Latex (moderate emissions)
- Natural paints (lowest emissions)
- Milk paint (minimal emissions)

Drywall

It's not just a blank canvas:

Can grow mold if it gets wet due to:

- Paper backing providing food for mold
- Porous structure trapping moisture
- Lack of air circulation behind walls
- Condensation on cold surfaces

Sometimes holds recycled materials that may release:

- Heavy metals
- Industrial chemicals
- Sulfur compounds

- Various organic compounds

Can trap pollutants and release them later through:

- Absorption into gypsum
- Binding to paper surfaces
- Accumulation in wall cavities
- Temperature-induced off-gassing

Modern "green" drywall may contain:

- Antimicrobial treatments
- Moisture-resistant additives
- Alternative binding materials
- Synthetic gypsum from industrial processes

Flooring

What you walk on matters:

Vinyl flooring releases:

- Phthalates (plasticizers)
- VOCs from adhesives
- Cleaning chemical residues
- Wear particles from foot traffic

New carpet off-gasses dozens of chemicals including:

- 4-Phenylcyclohexene (new carpet smell)
- Styrene from backing
- Adhesive components
- Stain-resistant treatments

- Antimicrobial additives

Even hardwood floors can be problematic due to:

- Polyurethane finishes
- Stains and sealers
- Adhesives in engineered wood
- Cleaning product residues

Different flooring types have different issues:

- Laminate (formaldehyde from adhesives)
- Cork (sealant chemicals)
- Bamboo (binding agents)
- Tile (grout sealers and adhesives)

The problem is made worse by how airtight modern homes are. Back in the day, houses were pretty drafty – not great for energy bills, but at least there was plenty of air exchange with the outdoors. Modern homes are sealed up tight to save energy, which means pollutants can build up inside with nowhere to go. Let's look at how modern construction affects air quality:

Air Exchange Rates:

- Old homes: 1-2 air changes per hour naturally
- Modern homes: 0.1-0.2 air changes per hour
- Recommended rate: 0.35 air changes per hour minimum

Impact of reduced ventilation:

- Higher pollutant concentrations

- Increased humidity
- More biological growth
- Greater exposure to off-gassing

Energy-Efficient Features That Can Affect Air Quality:

- Weather stripping
- Double-pane windows
- Vapor barriers
- Spray foam insulation
- Air sealing techniques
- Modern HVAC systems

This brings us to something called "off-gassing." It's a term for when solid materials release chemicals into the air, and it's happening all around you right now. That "new" smell from furniture, carpets, or building materials? That's off-gassing in action. And it can continue long after the smell is gone.

Understanding Off-Gassing:

Primary off gassing occurs when:

- Materials are new
- Temperature increases
- Humidity rises
- Materials are disturbed

Secondary off-gassing happens through:

- Chemical breakdown over time

- Reactions with other pollutants
- UV exposure
- Physical wear and tear

Factors affecting off-gassing rates:

- Temperature (higher = faster)
- Humidity (higher = faster)
- Air flow (more = faster)
- Surface area exposed
- Age of material

Let's talk about cleaning products – they're supposed to make our homes cleaner, right? Well, they might be making your surfaces sparkle while making your air dirtier. Many common cleaning products release a whole cocktail of chemicals into the air:

All-purpose cleaners contain:

Ammonia which can:

- Irritate eyes and lungs
- React with other chemicals
- Create persistent odors
- Trigger asthma attacks

Ethylene glycol ethers that can:

- Cause reproductive issues
- Affect blood formation
- Damage liver and kidneys

- Cause neurological effects

Artificial fragrances containing:

- Phthalates
- Synthetic musks
- Allergens
- Masking agents

Window cleaners release:

- Ammonia compounds

Isopropyl alcohol that can:

- Irritate eyes and respiratory tract
- Cause headaches
- Create dizziness
- Affect central nervous system

Fragrances with:

- Unknown chemical combinations
- Potential allergens
- Hormone disruptors
- Long-lasting residues

Oven cleaners emit:

Sodium hydroxide (lye) which can:

- Cause severe respiratory irritation
- Create caustic fumes

- React with other chemicals
- Persist in indoor air

Ethylene glycol with:

- Sweet-smelling but toxic fumes
- Long-term health effects
- Ability to penetrate skin
- Cumulative exposure risks

Various solvents including:

- Methylene chloride
- Acetone
- Butyl compounds
- Glycol ethers

Now let's get really gross and talk about biological pollutants. Your home is literally alive with microscopic organisms and biological materials, forming a complex indoor ecosystem that scientists call the "indoor microbiome." Let's break down these invisible roommates:

Dust Mites:

Population density:

- Up to 19,000 mites per gram of dust
- Higher numbers in humid climates
- More abundant in bedding and furniture
- Increase with human occupation

Feeding habits:

- Consume dead skin cells (we shed about 1.5 grams daily)
- Feed on fungi and bacteria
- Thrive on pet dander
- Need humidity above 50% to survive

Allergen production:

- Create at least 15 different allergens
- Proteins in fecal pellets are most allergenic
- Dead mites continue releasing allergens
- Allergens can remain airborne for hours

Control methods:

- Regular washing in hot water (130°F/54°C)
- Maintaining humidity below 50%
- Using allergen-proof covers
- Regular vacuum cleaning with HEPA filters

Mold:

Common indoor species:

- Aspergillus (can produce mycotoxins)
- Penicillium (some species produce toxins)
- Stachybotrys (infamous "black mold")
- Alternaria (common allergen)
- Cladosporium (can grow at lower temperatures)

Growth conditions:

- Needs moisture content above 60%
- Grows on most building materials
- Thrives in temperatures 40-100°F (4-38°C)
- Can grow within 24-48 hours of water damage

Health effects from:

- Spores (reproductive structures)
- Mycotoxins (toxic compounds)
- MVOCs (microbial volatile organic compounds)
- Glucans (cell wall components)
- Allergens (protein compounds)

Hidden growth locations:

- Behind wallpaper
- Inside wall cavities
- Under carpeting
- In HVAC systems
- Above ceiling tiles

Pet Dander:

Characteristics:

- Microscopic skin flakes
- Contains specific proteins that trigger allergies
- Can remain airborne for hours
- Sticky enough to attach to all surfaces

Distribution patterns:

- Spreads through entire building
- Concentrates in soft furnishings
- Travels on clothing
- Enters buildings even where no pets live

Persistence in environment:

- Can last months after pet removal
- Remains allergenic for extended periods
- Difficult to eliminate
- Builds up in ventilation systems

Control strategies:

- Regular pet grooming
- HEPA air filtration
- Frequent cleaning of surfaces
- Keeping pets out of bedrooms

Bacteria:

Indoor sources:

- Human occupants (we each release about 37 million bacteria per hour)
- Pet presence
- Water damage
- HVAC systems
- Soil brought in on shoes

Common types:

- Staphylococcus (skin-associated)
- Streptococcus (respiratory associated)
- Pseudomonas (water-associated)
- Bacillus (soil-associated)
- Legionella (water system-associated)

Growth conditions:

- Moisture requirements vary by species
- Many thrive at room temperature
- Some need extremely specific conditions
- Can form biofilms on surfaces

Health impacts:

- Direct infection risks
- Endotoxin production
- Allergen creation
- Odor generation
- Interaction with other pollutants

Viruses:

Transmission methods:

- Airborne droplets
- Surface contact
- HVAC systems
- Human-to-human spread

Survival factors:

- Humidity levels (usually prefer lower)
- Temperature
- Surface type
- UV exposure

Common indoor viruses:

- Rhinovirus (common cold)
- Influenza
- Coronavirus
- Norovirus
- Respiratory syncytial virus (RSV)

Control measures:

- Ventilation
- Air filtration
- UV germicidal irradiation
- Surface disinfection
- Humidity control

These biological pollutants can be especially problematic because they can reproduce. Unlike chemical pollutants that eventually break down or dissipate, biological pollutants can increase over time if conditions are right. Let's look at what makes them thrive:

Environmental Conditions Supporting Growth:

Moisture:

- Relative humidity above 60%
- Water leaks
- Condensation
- Poor ventilation
- Humidifier use

Temperature:

- Most prefer 68-86°F (20-30°C)
- Some can grow at lower temperatures
- Growth rates increase with warmth
- Temperature fluctuations can cause condensation

Food sources:

- Dust
- Skin cells
- Building materials
- Pet food
- Organic debris

Speaking of buildings that make us sick, let's talk about "Sick Building Syndrome" (SBS) in detail. It's a real condition where people experience health problems that seem to be linked to time spent in a particular building, but no specific cause can be identified.

SBS Characteristics:

Temporal patterns:

- Symptoms appear after entering building
- Worsen throughout the day
- Improve after leaving
- Return upon re-entry

Affected populations:

- More common in office workers
- Higher incidence in women
- Increased risk in certain jobs
- Variable individual susceptibility

Symptoms of SBS can include:

Respiratory:

- Nose irritation and congestion
- Throat irritation
- Chest tightness
- Coughing
- Wheezing

Neurological:

- Headaches
- Difficulty concentrating
- Mental fatigue
- Memory problems
- Dizziness

Dermatological:

- Skin dryness
- Itching
- Rashes
- Unusual sensitivity

General:

- Fatigue
- Nausea
- Eye irritation
- Muscle pain
- General malaise

Several factors can contribute to SBS:

Physical Building Factors:

Ventilation:

- Insufficient fresh air intake
- Poor air distribution
- Contaminated ventilation systems
- Blocked vents
- Imbalanced air pressure

Temperature and Humidity:

- Poor temperature control
- Humidity extremes
- Temperature stratification
- Cold/hot spots

- Radiant heat/cold surfaces

Lighting:

- Insufficient natural light
- Glare
- Flicker from fluorescent lights
- Poor color rendering
- Uneven distribution

Noise:

- HVAC system noise
- Equipment vibration
- Poor acoustics
- External noise intrusion
- Low-frequency noise

Chemical Factors:

Indoor sources:

- Building materials
- Office equipment
- Cleaning products
- Personal care products
- Furnishings

Outdoor sources:

- Vehicle exhaust
- Industrial emissions

- Construction activities
- Agricultural spraying
- Waste treatment facilities

Biological Factors:

Microbial growth:

- Mold in building materials
- Bacteria in HVAC systems
- Biofilm in humidifiers
- Contaminated ductwork
- Water-damaged materials

Human factors:

- Occupant density
- Activity levels
- Personal hygiene
- Health status
- Stress levels

Modern office buildings are particularly prone to SBS because they often:

- Have sealed windows that:
- Prevent natural ventilation
- Create dependency on mechanical systems
- Isolate occupants from outside air
- Increase energy efficiency but reduce air exchange

Rely entirely on mechanical ventilation:

- Complex HVAC systems
- Potential for contamination
- Maintenance challenges
- Energy conservation conflicts

Contain lots of synthetic materials:

- Carpet
- Furniture
- Wall coverings
- Electronics
- Building materials

House many people and electronic equipment:

- High occupant density
- Numerous computers and printers
- Photocopiers
- Break room appliances
- Personal devices

Use harsh cleaning products:

- Industrial strength cleaners
- Disinfectants
- Floor strippers and waxes
- Air fresheners
- Pest control chemicals

But here's some good news: there are many specific things you can do to improve your indoor air quality. Let's break down these solutions into comprehensive, actionable strategies:

Ventilation - Your First Line of Defense:

Natural Ventilation Strategies:

- Cross ventilation (opening windows on opposite sides)
- Stack effect ventilation (using temperature differences)
- Window timing (early morning/evening when pollution is lowest)
- Season-specific approaches (spring/fall versus summer/winter)
- Creating airflow patterns

Mechanical Ventilation Options:

- Exhaust fans rated for room size
- Heat Recovery Ventilators (HRVs)
- Energy Recovery Ventilators (ERVs)
- Whole-house fans
- Fresh air intake systems

Ventilation Best Practices:

- Run bathroom fans for 20 minutes after showers
- Use kitchen range hoods that vent outside
- Clean ventilation components regularly
- Balance input and output airflow

- Monitor humidity levels

Source Control - Stop Pollution Before It Starts:

Product Selection:

- Low-VOC or zero-VOC materials
- Paint and finishes
- Adhesives and sealants
- Cleaning products
- Furniture
- Building materials
- Natural alternatives to synthetic products
- Beeswax candles instead of scented
- Essential oils instead of air fresheners
- Vinegar and baking soda for cleaning
- Wool and cotton instead of synthetic fabrics
- Natural wood instead of pressed wood

Moisture Control:

Fix leaks immediately:

- Plumbing
- Roof
- Windows
- Foundation
- Appliance connections

Manage humidity:

- Use dehumidifiers in damp areas

- Monitor relative humidity (keep between 30-50%)
- Address condensation issues
- Proper bathroom and kitchen ventilation
- Regular maintenance of HVAC systems

Chemical Storage:

Keep chemicals out of living spaces:

- Store in detached garage or shed
- Use airtight containers
- Check for leaks regularly
- Dispose of unnecessary products
- Follow storage temperature guidelines

Safe handling practices:

- Use only in well-ventilated areas
- Follow manufacturer instructions
- Wear appropriate protective equipment
- Never mix products
- Keep products in original containers

Air Cleaning - Active Pollution Removal:

HVAC Filtration:

Filter selection:

- MERV ratings explained (higher is better, but check system compatibility)
- HEPA vs. regular filters

- Electrostatic options
- Washable vs. disposable
- Size and fit considerations

Maintenance schedule:

- Check monthly
- Replace every 3-6 months
- Clean reusable filters
- Inspect ductwork
- Monitor system performance

Portable Air Purifiers:

Technology types:

- HEPA filtration
- Activated carbon
- UV-C light
- Ionization
- Photocatalytic oxidation

Selection criteria:

- Room size matching
- Clean Air Delivery Rate (CADR)
- Noise levels
- Energy efficiency
- Maintenance requirements

Placement strategies:

- Near pollution sources
- Away from obstacles
- In frequently used spaces
- Consider air flow patterns
- Multiple units for large spaces

Natural Air Cleaning:

Air-purifying plants:

- Snake plant
- Spider plant
- Peace lily
- Boston fern
- Bamboo palm

Placement considerations:

- Light requirements
- Humidity needs
- Space availability
- Maintenance access
- Pet safety

Maintenance - Regular Upkeep for Clean Air:

Cleaning Strategies:

Daily tasks:

- Vacuum with HEPA filter

- Dust with microfiber cloths
- Remove shoes at entry
- Empty pet litter
- Wipe down often touched surfaces

Weekly tasks:

- Wash bedding in hot water
- Clean all floors
- Check for moisture issues
- Clean pet areas
- Inspect air vents

Monthly tasks:

- Deep clean carpets
- Check HVAC filters
- Clean ventilation fans
- Inspect for mold
- Clean air purifiers

HVAC Maintenance:

Professional service:

- Annual inspections
- Duct cleaning when needed
- Coil cleaning
- System balancing
- Performance testing

DIY maintenance:

- Filter changes
- Vent cleaning
- Condensate drain checks
- Thermostat battery replacement
- Register cleaning

Monitoring and Testing:

Air Quality Monitoring:

Parameters to track:

- Temperature
- Humidity
- Carbon dioxide
- VOCs
- Particulate matter

Monitoring devices:

- Simple humidity meters
- CO2 monitors
- VOC sensors
- Particle counters
- Professional testing equipment

Professional Assessment:

When to get help:

- Persistent symptoms

- After water damage
- Before/after renovations
- New building occupancy
- Suspected contamination

Types of testing:

- Mold testing
- VOC analysis
- Radon testing
- Asbestos inspection
- Lead testing

Long-term Strategies:

Building Improvements:

Structural upgrades:

- Better insulation
- Window replacement
- Vapor barriers
- Proper drainage
- Radon mitigation

System upgrades:

- Advanced HVAC systems
- Whole-house filtration
- UV air treatment
- Smart ventilation controls

- Energy recovery systems

Lifestyle Changes:

Daily habits:

- No smoking indoors
- Regular cleaning
- Proper ventilation use
- Chemical product reduction
- Moisture management

Long-term practices:

- Green cleaning adoption
- Natural product transition
- Regular maintenance routines
- Air quality awareness
- Health monitoring

Remember that improving indoor air quality is often about balance. While you want to reduce pollutants, you also need to maintain comfortable temperature and humidity levels. Sometimes this means making trade-offs, like choosing when to open windows based on outdoor air quality and weather conditions.

Consider keeping a "pollution diary" if you're concerned about indoor air quality. Note when symptoms occur and what might have triggered them.

This can help you find problem areas and make targeted improvements. Include:

- Date and time of symptoms
- Location in the building
- Recent activities
- Weather conditions
- Building conditions
- Recent changes or events

Think of managing indoor air quality as an ongoing process rather than a one-time fix. It requires:

- Regular monitoring
- Consistent maintenance
- Prompt problem response
- Continuous improvement
- Occupant awareness

In the next chapter, we'll look at specific health effects of poor air quality and how different pollutants affect different people. But for now, look around your home with fresh eyes. What potential sources of indoor air pollution can you spot? What simple changes could you make today to start breathing cleaner air?

Remember, while the challenge of indoor air pollution might seem overwhelming, every small improvement helps. Start with the easiest changes and work your way up to more complex solutions. Your lungs (and your health) will thank you for it!

Because when it comes to indoor air quality, knowledge really is power – the power to make better choices, create healthier spaces, and breathe easier in your own home. After all, home should be where you can take a deep, clean breath and truly relax.

CHAPTER 6

IMMEDIATE HEALTH EFFECTS

Let's have an honest conversation about something we'd probably rather not think about what happens to our bodies when we breathe bad air. You know that scratchy feeling in your throat on a smoggy day? Or that tight chest sensation when you walk into a freshly painted room? Those aren't just annoying sensations – they're your body's alarm system telling you something's not right with the air you're breathing.

Think of your body like a sophisticated air quality monitor. Before any fancy equipment can detect problems, your body is already sending you signals. The trick is learning to recognize these signals and understand what they're telling you. It's like learning a new language – the language of your body's response to air pollution.

And let me tell you, your body speaks this language fluently! It's been perfecting this communication system for millions of years of evolution. Long before we had air quality sensors and pollution alerts, our ancestors relied on these bodily signals to avoid dangerous air. That coughing reflex? It's been protecting lungs since mammals first crawled out of the sea. That urge to move away from irritating smoke. It's an ancient survival mechanism that's still serving us today.

Let's start with your respiratory system, because that's air pollution's first stop on its unwelcome tour of your body. Your respiratory system is like an amazing air-processing plant, and it's got some impressive security measures:

Your nose is the first line of defense:

- Nasal hairs (cilia) trap larger particles
- Mucus membranes catch smaller particles
- Warm and humidify incoming air
- Has nerve endings that detect irritants
- Can trigger protective sneezing reflexes

Your throat and large airways are like security checkpoints:

- Lined with sticky mucus to trap particles
- Equipped with hair-like cilia that move debris up and out
- Have smooth muscles that can tighten to restrict access
- Contain inflammatory cells ready to fight invaders
- Can trigger protective coughing reflexes

Your bronchial tubes form an intricate network:

- Branch into smaller and smaller airways
- Have muscles that can constrict to limit exposure
- Lined with specialized cells that produce mucus

- Contains immune cells ready to respond
- Can signal inflammation when threatened

Your tiny air sacs (alveoli) are the final frontier:

- Extremely thin walls for gas exchange
- Covered with surfactant to maintain surface tension
- Protected by specialized immune cells
- Extremely vulnerable to damage
- Critical for oxygen delivery to blood

When you breathe polluted air, this system immediately goes into defensive mode, kind of like a building going into lockdown:

Your nose might start running to flush out irritants through:

- Increased mucus production
- Enhanced blood flow to the area
- Activation of nerve endings
- Triggered sneeze reflexes
- Release of inflammatory mediators

Your throat might feel scratchy as it:

- Increases mucus production
- Triggers cough reflexes
- Activates inflammatory responses
- Tightens muscles
- Signals pain receptors

Your airways might tighten to protect your lungs by:

- Bronchial muscle constriction
- Increased mucus production
- Inflammatory cell activation
- Blood vessel dilation
- Nerve ending stimulation

You might start coughing to expel unwanted substances through:

- Forceful air expulsion
- Mucus clearance
- Muscle coordination
- Reflex activation
- Protective spasms

Here's what's happening with some common respiratory responses, and trust me, it's like watching a sophisticated defense system in action:

Coughing:

Types of coughs:

- Dry (non-productive) coughs from irritation
- Wet coughs clearing mucus
- Spasmodic coughs from reflexes
- Chronic coughs from ongoing exposure

Timing patterns:

- Immediate response to irritants
- Delayed reaction after exposure
- Night-time worsening
- Exercise-induced episodes

Associated symptoms:

- Throat irritation
- Chest tightness
- Voice changes
- Muscle soreness

Wheezing:

Characteristics:

- High-pitched whistling sound
- Usually worse on exhaling
- Can be heard without stethoscope
- May change with position

Patterns:

- Intermittent or continuous
- Worse at night
- Exercise-induced
- Weather-dependent

Warning signs:

- Increasing severity

- Associated breathlessness
- Color changes in lips/skin
- Inability to speak in full sentences

Shortness of breath (dyspnea):

Sensations:

- Air hunger
- Chest tightness
- Unable to get deep breath
- Breathing requires effort

Triggers:

- Physical activity
- Lying flat
- Emotional stress
- Environmental conditions

Warning signs:

- Rapid breathing
- Use of accessory muscles
- Blue lips or fingertips
- Mental status changes

For people with asthma or other respiratory conditions, these effects can be particularly severe. It's like their respiratory system's alarm is already set to sensitive, and air pollution just cranks up the volume. An asthma attack triggered by air pollution can be scary and dangerous:

Initial phase:

- Airways begin to spasm
- Mucus production increases
- Inflammation ramps up
- Breathing becomes labored
- Anxiety often begins

Peak phase:

- Airways severely constrict
- Breathing becomes difficult
- Panic often sets in
- Rescue medication needed
- Oxygen levels may drop

Recovery phase:

- Airways slowly relax
- Breathing improves gradually
- Fatigue sets in
- Residual sensitivity remains
- Monitoring needed

But your respiratory system isn't the only thing affected by air pollution. Let's talk about your heart and blood vessels. You might be wondering, "Wait, what does air pollution have to do with my heart?" Quite a lot, actually. When you breathe polluted air, your cardiovascular system responds in several ways:

Blood pressure changes occur because:

Immediate effects:

- Blood vessels constrict
- Stress hormones surge
- Heart rate increases
- Blood becomes stickier
- Inflammation begins

Delayed responses:

- Sustained elevation
- Variable patterns
- Nocturnal changes
- Exercise intolerance

Heart rhythm disturbances can develop:

Types of irregularities:

- Extra beats
- Skipped beats
- Rapid heartbeat
- Irregular patterns

Contributing factors:

- Oxygen reduction
- Stress response
- Inflammatory effects
- Neural disruption

Blood chemistry changes include:

Immediate changes:

- Increased inflammatory markers
- Enhanced clotting factors
- Stress hormone elevation
- Reduced oxygen saturation

Lasting effects:

- Persistent inflammation
- Altered clotting profile
- Metabolic changes
- Immune system activation

For people with heart conditions, these effects can be particularly dangerous. It's like adding stress to an already stressed system. Even short-term exposure to air pollution can trigger:

Chest Pain (Angina):

Characteristics:

- Pressure or squeezing sensation
- Radiation to arm, neck, or jaw
- Associated shortness of breath
- May come with sweating
- Triggered by minimal exertion

Warning signs:

- Increasing frequency
- Greater intensity
- Longer duration
- Less responsive to medication
- New patterns of pain

Heart Attacks:

Risk factors with pollution:

- Increased inflammation
- Enhanced blood clotting
- Reduced oxygen delivery
- Stressed heart muscle
- Electrical disruption

Warning symptoms:

- Chest discomfort
- Upper body pain
- Shortness of breath
- Cold sweats
- Unusual fatigue

Stroke:

Mechanisms:

- Blood clot formation
- Vessel inflammation

- Blood pressure spikes
- Rhythm disturbances
- Reduced blood flow

Warning signs:

- Sudden weakness
- Speech problems
- Vision changes
- Balance issues
- Severe headache

Now, let's talk about something that might surprise you – air pollution's effects on your brain. Yes, your brain! Those tiny pollution particles don't just stay in your lungs; some can actually get into your bloodstream and affect your nervous system. Think of it as unwanted visitors crossing the blood-brain barrier.

The blood-brain barrier is like your brain's security system, but some pollutants are sneaky enough to get past it:

- Ultrafine particles can slip through
- Inflammatory signals can cross
- Chemical messengers get disrupted
- Nerve function gets affected
- Brain chemistry changes

Common neurological symptoms include:

Headaches:

Types:

- Tension headaches
- Migraine-like pain
- Pressure headaches
- Sinus-related pain
- Mixed pattern headaches

Characteristics:

- Location (where it hurts)
- Intensity (how bad it hurts)
- Duration (how long it lasts)
- Pattern (constant or pulsing)
- Associated symptoms

Triggers:

- Specific pollutants
- Length of exposure
- Activity levels
- Environmental conditions
- Individual sensitivity

Dizziness:

Types:

- Vertigo (spinning sensation)

- Light-headedness
- Balance problems
- Spatial disorientation
- Floating sensation

Associated symptoms:

- Nausea
- Vision problems
- Coordination issues
- Anxiety
- Weakness

Impact on activities:

- Walking difficulties
- Driving concerns
- Work performance
- Safety risks
- Social limitations

Mental fog:

Cognitive effects:

- Reduced attention span
- Memory problems
- Slower processing speed
- Decision-making difficulties
- Learning challenges

Work impact:

- Reduced productivity
- Increased errors
- Communication problems
- Safety concerns
- Job performance issues

Daily life effects:

- Task completion problems
- Social interaction difficulties
- Driving challenges
- Educational impacts
- Quality of life reduction

Allergic reactions are another common immediate response to air pollution. Your immune system sees certain pollutants as invaders and launches its defense system. This can trigger a cascade of reactions:

Nasal symptoms:

Immediate responses:

- Sneezing fits
- Runny nose
- Nasal congestion
- Post-nasal drip
- Itchy nose

Secondary effects:

- Sinus pressure
- Voice changes
- Sleep disruption
- Taste changes
- Ear pressure

Complications:

- Sinus infections
- Nose bleeds
- Chronic inflammation
- Polyp formation
- Structural changes

Eye reactions:

Surface effects:

- Redness
- Tearing
- Burning
- Itching
- Foreign body sensation

Vision impacts:

- Blurred vision
- Light sensitivity
- Focusing difficulties
- Reduced clarity

* Eye strain

Protective responses:

* Increased blinking
* Tear production
* Squinting
* Eye rubbing
* Eyelid swelling

Skin responses:

Direct reactions:

* Redness
* Itching
* Burning
* Rashes
* Hives

Pattern variations:

* Localized reactions
* Widespread responses
* Contact patterns
* Delayed reactions
* Chronic changes

Contributing factors:

* Sweating
* Temperature

- Humidity
- Physical contact
- Pre-existing conditions

Emergency Response Planning:

Immediate Response to Exposure:

Mild symptoms:

Recognition steps:

- Name potential sources
- Monitor symptom progression
- Document timing and patterns
- Note environmental conditions
- Track response to changes

Action steps:

- Remove from exposure
- Seek fresh air
- Monitor symptoms
- Document experience
- Prepare for recurrence

Moderate symptoms:

Assessment:

- Symptom severity
- Rate of progression

- Response to initial measures
- Available resources
- Support needs

Response:

- Use prescribed medications
- Implement protection measures
- Seek medical advice
- Monitor vital signs
- Plan follow-up care

Severe symptoms:

Emergency signs:

- Breathing difficulty
- Chest pain
- Mental status changes
- Severe dizziness
- Visual disturbances

Critical actions:

- Call emergency services
- Use emergency medications
- Document exposure details
- Maintain open airway
- Monitor consciousness

Prevention Strategies:

Indoor Environments:

Workplace:

- Ventilation checks
- Air quality monitoring
- Protection equipment
- Emergency procedures
- Exposure documentation

Home:

- Source control
- Ventilation management
- Filtration systems
- Humidity control
- Regular maintenance

Outdoor Activities:

Planning:

- Check air quality forecasts
- Time activities appropriately
- Choose suitable locations
- Prepare emergency supplies
- Have backup plans

Adaptation:

- Modify activity levels

- Use protective equipment
- Monitor symptoms
- Stay hydrated
- Know your limits

Special Population Considerations:

Children:

Unique vulnerabilities:

- Developing respiratory system
- Higher breathing rates
- More outdoor activity
- Hand-to-mouth behavior
- Limited communication skills

Protection strategies:

- Indoor activity during poor air quality
- Regular health monitoring
- Symptom education
- Emergency plan awareness
- Communication with caregivers

Elderly:

Special concerns:

- Reduced respiratory capacity
- Multiple health conditions
- Medication interactions

- Delayed symptoms recognition
- Limited mobility

Support needs:

- Regular monitoring
- Emergency planning
- Communication systems
- Access to care
- Support network

Ongoing Monitoring:

Symptom Documentation:

What to record:

- Exposure details
- Symptom progression
- Response to interventions
- Recovery patterns
- Long-term effects

Documentation methods:

- Written logs
- Digital tracking
- Photo evidence
- Medical records
- Environmental data

Pattern Recognition:

Identifying triggers:

- Environmental conditions
- Activity patterns
- Time patterns
- Location factors
- Combined effects

Using the information:

- Prevention planning
- Treatment modification
- Activity adjustments
- Communication with healthcare providers
- Emergency preparation

Remember, your body's warning signals are like your personal air quality monitor. Learning to recognize and respond to these signals can help protect you from more serious health effects. Don't ignore these warnings – they're your body's way of saying "Hey, something's not right here!"

In our next chapter, we'll look at the long-term health effects of air pollution exposure. But for now, pay attention to how your body responds to different air quality situations. You might be surprised at what it's trying to tell you.

After all, when it comes to air pollution, your body often knows best. Those immediate reactions aren't just inconvenient symptoms – they're important warning signs that deserve our attention and respect. By understanding and responding to these signals, we can better protect our health in both the short and long term.

Think of it as developing a partnership with your body's warning system. The more you pay attention to and understand these signals, the better equipped you'll be to protect yourself from air pollution's harmful effects. It's like having a built-in early warning system – you just need to learn its language and take its messages seriously.

CHAPTER 7

LONG-TERM HEALTH CONSEQUENCES

Think of air pollution like a really slow-motion disaster movie. While the immediate effects we discussed in the last chapter are like the dramatic scenes everyone notices, the long-term effects are more like the subtle plot points that build up over time. They might not be as obvious, but they're often more serious – and unfortunately, this isn't a movie we can just turn off.

Imagine dropping a single drop of food coloring into a glass of water every day. At first, you might not notice any change. But over time, that water's going to get darker and darker. That's kind of how long-term exposure to air pollution works in our bodies. Each exposure might seem minor, but they add up over time, potentially leading to serious health problems.

Think about it this way: your body is like a bank account, but instead of money, we're talking about health. Every exposure to air pollution is like a small withdrawal. One or two won't bankrupt you, but years of constant withdrawals. That's when you start seeing some serious negative balances in your health account.

Let's start with chronic diseases – the uninvited house guests that move in slowly and don't want to leave. Breathing polluted air day after day can lead to several chronic conditions:

Respiratory diseases come in various forms, each with its own special brand of misery:

Chronic Obstructive Pulmonary Disease (COPD):

- Progressive damage to airways
- Irreversible lung changes
- Reduced oxygen exchange
- Increased infection risk
- Quality of life impact

Treatment challenges:

- Daily medication needs
- Regular medical monitoring
- Lifestyle modifications
- Oxygen therapy potential
- Emergency plan necessity

Chronic bronchitis presents as:

- Persistent inflammation
- Excessive mucus production
- Daily cough for months
- Frequent infections
- Breathing difficulties

Long-term effects:

- Airway remodeling
- Muscle weakening
- Reduced exercise tolerance
- Social life limitations
- Work capacity reduction

Emphysema develops through:

- Alveoli destruction
- Reduced lung elasticity
- Air trapping in lungs
- Progressive breathlessness
- Chest expansion

Daily impacts:

- Difficulty with simple tasks
- Energy conservation needs
- Breathing strategy requirements
- Nutritional challenges
- Anxiety management

Think about your lungs like the air filter in your car. Just as a car's filter gets clogged and dirty over time if you drive on dusty roads, your lungs can get damaged from years of breathing polluted air. The difference is you can't just swap out your lungs for new ones.

Cardiovascular diseases are another major concern. Remember how we talked about air pollution affecting your heart in the short term? Well, those effects can become permanent over time:

Arterial Problems:

Hardening of arteries through:

- Plaque buildup
- Inflammation
- Wall thickening
- Reduced flexibility
- Compromised blood flow

Progressive damage leading to:

- Reduced organ perfusion
- Increased heart workload
- Tissue damage risk
- Clot formation potential
- Emergency situation risk

Blood Pressure Issues:

Chronic elevation causing:

- Blood vessel damage
- Heart enlargement
- Kidney stress
- Brain risks
- Vision problems

Management challenges:

- Medication needs
- Lifestyle changes
- Regular monitoring
- Dietary restrictions
- Exercise modifications

Heart Disease Manifestations:

Structural changes:

- Chamber enlargement
- Wall thickening
- Valve problems
- Rhythm disturbances
- Muscle weakening

Functional impacts:

- Reduced pumping efficiency
- Exercise intolerance
- Fluid retention
- Energy reduction
- Sleep disruption

It's like your cardiovascular system is running a marathon, but instead of getting stronger, it's just getting worn down because the air is full of hurdles it wasn't designed to handle.

Now let's talk about something that nobody wants to discuss but everyone needs to know about – cancer risks. Yes, certain air pollutants can increase your risk of developing several types of cancer:

Lung Cancer Development:

Risk factors:

- Particle exposure
- Chemical carcinogens
- Duration of exposure
- Concentration levels
- Individual susceptibility

Detection challenges:

- Late symptom onset
- Screening limitations
- Diagnostic delays
- Treatment complexity
- Poor prognosis potential

Other Cancer Connections:

Bladder cancer links:

- Industrial exposure correlation
- Chemical absorption
- Cellular changes
- DNA damage
- Progressive development

Liver cancer factors:

- Toxin accumulation
- Metabolic stress
- Inflammatory response
- Cellular mutations
- Organ stress

The relationship between air pollution and cancer is like playing a very unfair game of chance. Exposure doesn't guarantee you'll get cancer, but it can definitely stack the deck against you.

Let's move on to something that often gets overlooked – reproductive health. Air pollution can affect both male and female reproductive systems:

Female Reproductive Impacts:

Fertility effects:

- Egg quality reduction
- Hormonal disruption
- Implantation difficulties
- Cycle irregularities
- Conception challenges

Pregnancy risks:

- Placental problems
- Fetal growth restriction
- Blood pressure issues

- Gestational diabetes
- Preterm labor

Male Reproductive Concerns:

Sperm health:

- Count reduction
- Motility problems
- Morphology changes
- DNA fragmentation
- Fertility reduction

Hormonal impacts:

- Testosterone levels
- Endocrine disruption
- Reproductive drive
- Secondary characteristics
- Developmental issues

It's like trying to grow a garden in contaminated soil – even if everything else is perfect, the environmental conditions can make it much harder for life to flourish.

[Would you like me to continue with the remaining sections, including child development, genetic effects, healthcare costs, and societal impacts?]

Speaking of new life, let's talk about how air pollution affects child development. Kids are especially vulnerable because their bodies are still growing and developing:

Physical Development Impacts:

Respiratory system effects:

- Reduced lung capacity
- Airway hypersensitivity
- Increased infection risk
- Asthma development
- Chronic inflammation

Growth patterns:

- Height potential reduction
- Weight gain issues
- Bone density concerns
- Muscle development
- Organ maturation
- Immune system challenges
- Reduced resistance
- Increased allergies
- Autoimmune tendencies
- Frequent infections
- Recovery difficulties

Cognitive Development Concerns:

- Learning capacity
- Processing speed reduction
- Memory formation issues
- Attention span shortening

- Language development delays

  Problem-solving difficulties

Behavioral impacts:

- Increased irritability
- Impulse control issues
- Social interaction challenges
- Emotional regulation problems
- Sleep disturbances

Academic performance:

- Lower test scores
- Reading comprehension issues
- Mathematical processing delays
- Creative thinking barriers
- Educational achievement gaps

Think of a child's developing body like wet cement – whatever marks get made now can become permanent. That's why protecting kids from air pollution is so crucial.

Now for something really sci-fi sounding but very real – genetic effects. Air pollution can actually affect our DNA:

DNA Damage Mechanisms:

Direct damage:

- Strand breaks
- Base modifications
- Crosslinking problems

- Mutation induction
- Repair interference

Epigenetic changes:

- Gene expression alterations
- Methylation patterns
- Chromatin modifications
- Regulatory RNA impacts
- Inherited changes

Cellular Aging Effects:

- Telomere shortening
- Oxidative stress increase
- Mitochondrial damage
- Cell death acceleration
- Repair mechanism overwhelm

Generational Impacts:

Inherited modifications:

- Gene expression changes
- Susceptibility patterns
- Disease predisposition
- Developmental influences
- Metabolic alterations

Future generation effects:

- Cumulative damage potential

- Adaptation challenges
- Evolution implications
- Population health impacts
- Species vulnerability

It's like air pollution is a sneaky editor, making unwanted changes to your genetic instruction manual. These changes can potentially be passed on to future generations, which means today's air pollution could affect people who haven't even been born yet.

Let's talk dollars and cents for a moment, because the healthcare costs of air pollution are staggering:

Direct Medical Expenses:

Immediate care costs:

- Emergency room visits
- Hospital admissions
- Specialist consultations
- Diagnostic testing
- Medication expenses

Long-term care needs:

- Chronic disease management
- Rehabilitation services
- Medical equipment
- Home health care
- Preventive measures

Healthcare system burden:

- Facility demands
- Staff requirements
- Resource allocation
- Insurance impacts
- Budget strains

Indirect Economic Impact:

Productivity losses:

- Missed workdays
- Reduced efficiency
- Early retirement
- Career limitations
- Income reduction

Quality of life costs:

- Activity restrictions
- Lifestyle modifications
- Social limitations
- Emotional burden
- Family stress

Societal burden:

- Disability benefits
- Healthcare subsidies
- Support services
- Infrastructure needs

- Research funding

The societal impact goes even deeper:

Healthcare System Strain:

Resource allocation:

- Hospital bed occupation
- Staff time demands
- Equipment usage
- Medicine supplies
- Facility capacity

System adaptations:

- Emergency response plans
- Preventive programs
- Staff training
- Infrastructure updates
- Protocol development

Economic Ripple Effects:

Business impact:

- Worker productivity
- Insurance costs
- Facility maintenance
- Environmental compliance
- Legal liability

Property considerations:

- Value reduction
- Maintenance needs
- Insurance rates
- Development limitations
- Renovation requirements

Environmental Degradation:

Agricultural effects:

- Crop yield reduction
- Soil contamination
- Water quality impact
- Biodiversity loss
- Ecosystem disruption

Clean-up costs:

- Remediation efforts
- Prevention measures
- Monitoring systems
- Treatment facilities
- Research needs

It's like paying a hidden tax that nobody voted for, but everyone has to pay. The World Bank estimates that air pollution costs the global economy trillions of dollars every year.

But here's where it gets really interesting (and concerning) – these health effects don't hit everyone equally. Certain groups tend to bear more of the burden:

Vulnerable Population Impacts:

Low-income communities:

- Higher exposure levels
- Limited healthcare access
- Fewer protective resources
- Housing challenges
- Occupational risks

Children's vulnerability:

- Developmental impacts
- Long-term health effects
- Educational consequences
- Future potential limitation
- Quality of life reduction

Elderly concerns:

- Increased susceptibility
- Multiple health conditions
- Limited mobility
- Social isolation
- Care requirements

Geographical Disparities:

Urban concentrations:

- Traffic pollution
- Industrial exposure
- Population density
- Heat island effects
- Limited green space

Rural challenges:

- Agricultural chemicals
- Burning practices
- Limited monitoring
- Healthcare access
- Resource availability

The good news (yes, there is some!) is that our bodies have amazing healing capabilities. Studies have shown that when air quality improves, health outcomes get better:

Recovery Potential:

Respiratory improvement:

- Lung function gains
- Symptom reduction
- Exercise tolerance
- Infection resistance
- Quality of life

Cardiovascular benefits:

- Blood pressure normalization
- Heart rhythm improvement
- Vessel health
- Exercise capacity
- Risk reduction

Prevention Strategies:

Personal actions:

- Air quality monitoring
- Filtration systems
- Ventilation improvements
- Exposure reduction
- Health monitoring

Community efforts:

- Policy advocacy
- Green initiatives
- Public education
- Infrastructure improvements
- Resource sharing

Remember, knowledge is power. **Understanding these long-term health consequences helps us:**

- Make informed decisions
- Protect vulnerable populations
- Support effective policies

- Create healthier communities
- Plan for future generations

Because when it comes to air pollution and health, the choices we make today can affect us for years or even decades to come. It's like we're all co-authors of our future health story – and the decisions we make about air quality are some of the most important plot points.

In our next chapter, we'll explore specific strategies and solutions for protecting ourselves and our communities from these long-term health effects. But for now, remember that every breath matters, and every action we take to improve air quality is an investment in our collective future health.

CHAPTER 8

AT-RISK GROUPS

Not all lungs are created equal. That's not something you'll find written in any medical textbook, but it's a truth that becomes clear when we look at how air pollution affects diverse groups of people. Just like some people get sunburned more easily than others, some people are more vulnerable to air pollution's effects.

Think of it like a game of cards. We're all dealt different hands when it comes to how our bodies handle air pollution. Some people start with a full house of healthy lungs, strong immune systems, and good living conditions. Others might be playing with a pair of twos – dealing with asthma, living near pollution sources, or working in hazardous conditions. And just like in poker, the stakes are high – we're talking about people's health and lives.

Let's start with our smallest and most vulnerable players: children. Kids aren't just small adults – their bodies work differently in ways that make them especially susceptible to air pollution:

Physical Vulnerabilities:

Respiratory differences:

- Higher breathing rates relative to body size
- More air inhaled per pound
- Greater exposure to airborne pollutants
- Increased deposition of particles
- Enhanced absorption of toxins

Developmental factors:

- Growing lung tissue more vulnerable
- Developing immune system
- Rapid cell division period
- Incomplete defense mechanisms
- Ongoing organ development

Behavioral Factors:

Activity patterns:

- More time outdoors
- Higher activity levels
- Mouth breathing during play
- Ground-level exposure
- Hand-to-mouth behavior

Environmental exposure:

- School location impacts
- Playground proximity to traffic
- Sports field air quality

- Indoor school air issues
- Transportation exposure

Imagine building a house while someone's throwing dirt in your cement mixer. That's kind of what happens when kids' developing lungs are exposed to air pollution. The damage done during these crucial development years can affect them for life:

Long-term Impacts:

Respiratory development:

- Reduced maximum lung capacity
- Altered airway development
- Compromised breathing efficiency
- Increased susceptibility to infections
- Chronic inflammation patterns

Immune system effects:

- Modified immune responses
- Increased allergy risk
- Autoimmune tendencies
- Reduced infection resistance
- Chronic health vulnerabilities

Educational Impacts:

Academic performance:

- Concentration difficulties

- Memory problems
- Learning delays
- Behavioral issues
- Achievement gaps

Attendance patterns:

- Increased sick days
- Medical appointments
- Activity limitations
- Reduced participation
- Social isolation

On the other end of the age spectrum, we have our elderly population. Our seniors face their own set of challenges when it comes to air pollution:

Age-Related Vulnerabilities:

Physiological changes:

- Reduced lung elasticity
- Weakened respiratory muscles
- Compromised immune function
- Decreased heart efficiency
- Slower recovery ability

Compounding factors:

- Multiple health conditions
- Medication interactions
- Reduced mobility

- Social isolation
- Limited resources

Health Impact Magnification:

Cardiovascular effects:

- Increased heart strain
- Blood pressure instability
- Rhythm disturbances
- Reduced oxygen delivery
- Circulation problems

Respiratory complications:

- Breathing difficulty
- Increased infections
- Chronic condition exacerbation
- Recovery challenges
- Hospitalization risk

Now let's talk about a group that's actually breathing for two – pregnant women. When an expectant mother breathes polluted air, she's not just exposing herself but also her developing baby:

Pregnancy Complications:

Early pregnancy risks:

- Implantation issues
- Miscarriage potential

- Developmental disruption
- Placental problems
- Hormone disruption

Later pregnancy issues:

- Growth restrictions
- Preterm labor risk
- Preeclampsia potential
- Gestational diabetes
- Delivery complications

Fetal Development Impacts:

Organ system effects:

- Brain development
- Lung maturation
- Heart formation
- Immune system development
- Metabolic programming

Long-term consequences:

- Childhood health issues
- Developmental delays
- Chronic disease risk
- Immune system alterations
- Cognitive impacts

People with pre-existing conditions are like canaries in the coal mine when it comes to air pollution. Their bodies often react more quickly and severely to poor air quality:

Asthma Sufferers:

Daily challenges:

- Trigger sensitivity
- Medication management
- Activity limitations
- Sleep disruption
- Anxiety management

Emergency situations:

- Attack frequency increase
- Severity escalation
- Rescue medication needs
- Hospital visits
- Recovery time

Quality of life impact:

- Work/school disruption
- Activity restrictions
- Social limitations
- Emotional stress
- Financial burden

Heart Disease Patients:

Cardiovascular stress:

- Blood pressure instability
- Rhythm irregularities
- Chest pain episodes
- Exercise intolerance
- Medication adjustments

Systemic effects:

- Inflammation increase
- Blood clot risk
- Oxygen demand
- Vessel constriction
- Stress response

Daily management:

- Activity modification
- Medication compliance
- Symptom monitoring
- Emergency planning
- Healthcare coordination

COPD Patients:

Respiratory challenges:

- Breathing difficulty
- Mucus production
- Infection risk

- Exercise limitation
- Sleep disruption

Disease progression:

- Accelerated decline
- Frequent exacerbations
- Hospital admissions
- Oxygen dependence
- Independence loss

Support needs:

- Medical monitoring
- Home modifications
- Caregiver assistance
- Equipment requirements
- Transportation help

Now let's talk about people who face occupational exposure to air pollution. These folks are like unwilling participants in a daily experiment with air quality:

Outdoor Workers:

Construction workers:

- Dust exposure
- Vehicle emissions
- Chemical fumes
- Weather factors
- Physical exertion

Traffic officers:

- Vehicle exhaust
- Particulate matter
- Extended exposure
- Weather challenges
- Limited protection

Agricultural workers:

- Pesticide exposure
- Dust inhalation
- Animal allergens
- Chemical sprays
- Seasonal variations

Indoor High-Risk Workers:

Factory workers:

- Chemical exposure
- Process emissions
- Confined spaces
- Ventilation issues
- Protection adequacy

Salon workers:

- Chemical fumes
- Product particles
- Confined space
- Chronic exposure

- Ventilation challenges

Cleaning professionals:

- Chemical mixtures
- Confined spaces
- Product interactions
- Duration exposure
- Protection limitations

Occupational Risk Factors:

Exposure patterns:

- Duration length
- Concentration levels
- Multiple pollutants
- Cumulative effects
- Recovery time

Workplace conditions:

- Ventilation adequacy
- Protection equipment
- Safety protocols
- Monitoring systems
- Emergency procedures

Environmental justice issues represent perhaps the most troubling aspect of air pollution vulnerability. It's not just about individual sensitivity – it's about systemic exposure patterns:

Low-Income Community Challenges:

Environmental factors:

- Industrial proximity
- Traffic concentration
- Poor infrastructure
- Limited green space
- Multiple pollution sources

Housing issues:

- Ventilation problems
- Maintenance deficits
- Mold presence
- Pest problems
- Structural issues

Resource limitations:

- Healthcare access
- Protective equipment
- Air purification
- Medical treatment
- Prevention measures

Minority Community Burdens:

Geographical factors:

- Industrial zoning
- Transportation corridors
- Urban heat islands

- Limited vegetation
- Emergency response gaps

Historical context:

- Residential segregation
- Infrastructure neglect
- Political marginalization
- Economic barriers
- Health disparities

Systemic challenges:

- Policy enforcement
- Resource allocation
- Monitoring coverage
- Remediation priority
- Community voice

Protection Strategies:

For Children:

School-based measures:

- Air quality monitoring
- Ventilation improvements
- Activity modifications
- Health education
- Emergency protocols

Community support:

- Clean air zones
- Traffic reduction
- Green space creation
- Healthcare access
- Education programs

For Elderly:

Individual protection:

- Air purification
- Activity planning
- Health monitoring
- Emergency preparation
- Support networks

Facility improvements:

- Ventilation systems
- Air quality monitoring
- Emergency protocols
- Staff training
- Resident education

For Pregnant Women:

Personal measures:

- Exposure avoidance
- Activity modification

- Indoor air quality
- Healthcare monitoring
- Support resources

Workplace accommodations:

- Remote work options
- Location changes
- Schedule flexibility
- Protection equipment
- Environmental controls

For Workers:

Workplace protections:

- Engineering controls
- Personal equipment
- Exposure monitoring
- Health surveillance
- Emergency procedures

Policy measures:

- Safety standards
- Enforcement
- Training requirements
- Medical monitoring
- Compensation systems

Community Action Steps:

Awareness building:

- Education programs
- Risk communication
- Resource sharing
- Support networks
- Advocacy training

Policy advocacy:

- Regulation enforcement
- Resource allocation
- Monitoring expansion
- Protection standards
- Community involvement

Remember, being at-risk doesn't mean being helpless. Knowledge, preparation, and support can make a stark difference:

Individual Empowerment:

Knowledge development:

- Risk understanding
- Protection methods
- Warning signs
- Resource access
- Support systems

Action planning:

- Emergency preparation
- Daily management
- Healthcare coordination
- Support networks
- Resource utilization

Community Support:

Network development:

- Neighbor awareness
- Resource sharing
- Emergency assistance
- Information exchange
- Advocacy cooperation

Resource allocation:

- Protection equipment
- Medical access
- Emergency services
- Support programs
- Education resources

In our next chapter, we'll look at specific solutions and strategies for protecting everyone from air pollution. But for now, take a moment to think about the at-risk people in your life and community. Because when it comes to air pollution, protecting our most vulnerable is protecting our future.

Remember, air pollution vulnerability isn't just about individual sensitivity – it's about systemic patterns of exposure and access to protection. By understanding these patterns, we can work toward solutions that protect everyone, especially those who need it most.

CHAPTER 9

MONITORING AND ASSESSMENT

Remember those mood rings that were popular back in the day? The ones that supposedly told you how you were feeling based on their color? Well, modern air quality monitoring is kind of like that – except instead of questionable jewelry readings, we've got sophisticated systems that actually tell us what's in our air. And trust me, this is one mood reading you'll want to pay attention to.

Let's start with how we keep track of what's floating around in our atmosphere. Imagine you're a detective trying to solve the mystery of what's in your air. You've got tools ranging from simple canaries in coal mines (yes, that was a real thing!) to modern sensors that can detect particles smaller than a human hair.

Speaking of those canaries, let's take a quick historical detour because it's fascinating stuff. Miners used to carry canaries underground because these little birds were super sensitive to toxic gases like carbon monoxide. If the canary stopped singing or, sadly, died, it was an early warning system telling miners to get out fast. We've come a long way from using birds as detectors, but the principle remains the same – we need early warning systems to protect us from dangerous air.

Today's air quality monitoring systems are like a neighborhood watch for pollution. They're constantly on patrol, checking for:

Particle Pollution:

PM2.5 (fine particles):

- Smoke particles
- Industrial emissions
- Vehicle exhaust
- Chemical reactions
- Organic compounds

PM10 (coarse particles):

- Dust
- Pollen
- Mold spores
- Construction debris
- Road particles

Gaseous Pollutants:

Ground-level ozone:

- Formation patterns
- Peak times
- Weather influences

- Precursor chemicals
- Seasonal variations

Carbon monoxide:

- Traffic patterns
- Industrial sources
- Indoor levels
- Weather effects
- Time variations

Chemical Compounds:

Sulfur dioxide:

- Industrial emissions
- Power plant output
- Volcanic activity
- Marine sources
- Chemical processes

Nitrogen dioxide:

- Vehicle emissions
- Power generation
- Industrial processes
- Urban concentrations
- Indoor sources

These monitoring stations are scattered throughout cities and rural areas like security cameras, keeping an eye on our air 24/7. But here's something interesting – the placement of these monitors is a science in itself. They need to be:

- High enough to avoid interference
- Away from immediate pollution sources
- Representative of the area
- Accessible for maintenance
- Secure from vandalism

Let's decode the Air Quality Index (AQI) – that colorful scale that goes from green (good) to purple (hide in your basement). It's like a report card for our air, but with some fascinating nuances:

Green Zone (0-50):

Characteristics:

- Clean, fresh air
- Good visibility
- Low particle counts
- Minimal health risk
- Ideal for outdoor activities

Health implications:

- Safe for everyone
- Great for exercise
- No restrictions needed
- Good for sensitive groups
- Perfect for children's activities

Yellow Zone (51-100):

Conditions:

- Moderate pollution
- Slightly reduced visibility
- Some particle presence
- Acceptable for most
- Watch for changes

Considerations:

- Monitor sensitive groups
- Plan activities accordingly
- Watch for worsening
- Check forecasts
- Stay informed

Orange Zone (101-150):

Situation:

- Unhealthy for sensitive groups
- Noticeable pollution
- Reduced visibility
- Potential health impacts
- Activity restrictions needed

Actions needed:

- Limit outdoor activity
- Monitor symptoms
- Indoor air cleaning

- Plan alternatives
- Health awareness

Red and Beyond (151+):

Serious conditions:

- Health alerts
- Visible pollution
- Significant risks
- Emergency measures
- Public warnings

Response requirements:

- Stay indoors
- Use air purification
- Follow health guidance
- Emergency planning
- Community response

But here's the thing about AQI readings – they're like weather forecasts for different neighborhoods. The air quality can vary dramatically even within a single city, which is why personal monitoring has become so important.

Speaking of personal monitors, let's dive deeper into these fascinating devices. Modern air quality monitors are like tiny laboratories you can carry around:

Basic Monitors:

Features:

- Particle counting
- Simple display
- Battery operation
- Portable design
- Basic trends

Limitations:

- Limited accuracy
- Few pollutants
- No data storage
- Basic metrics
- Simple analysis

Advanced Personal Devices:

Capabilities:

- Multiple pollutants
- Data logging
- Smartphone integration
- Historical tracking
- Pattern analysis

Advanced features:

- GPS tracking
- Weather integration
- Health correlations
- Alert systems
- Data sharing

Professional Equipment:

Laboratory grade:

High precision
Multiple parameters
Calibration features
Complex analysis
Research quality

Applications:

Research studies
Compliance monitoring
Health assessments
Environmental impact
Policy development

But all this monitoring equipment is useless without proper data interpretation. This is where science meets storytelling – taking all those numbers and turning them into information we can actually use. Think of it like being a translator for the language of air quality data:

Data Analysis Techniques:

Pattern Recognition:

Temporal trends:

- Daily cycles
- Weekly patterns
- Seasonal variations

- Annual trends
- Long-term changes

Spatial patterns:

- Urban-rural differences
- Hot spot identification
- Pollution gradients
- Geographic variations
- Source impacts

Statistical Analysis:

Basic metrics:

- Averages
- Peak values
- Minimum levels
- Standard deviations
- Trend lines

Advanced statistics:

- Correlation analysis
- Predictive modeling
- Source apportionment
- Health impact assessment
- Risk evaluation

Data Visualization:

Mapping techniques:

- Heat maps
- Pollution roses
- Contour plots
- Time series graphs
- Source attribution maps

Interactive tools:

- Real-time displays
- Web interfaces
- Mobile apps
- Alert systems
- Public dashboards

Global monitoring networks are like the United Nations of air quality – they bring together data from around the world. Here's how this massive system works:

Monitoring Infrastructure:

Ground-based systems:

Fixed stations:

- Urban monitors
- Rural background sites
- Industrial area coverage
- Traffic monitors

- Research stations

Mobile platforms:

- Research vehicles
- Drone systems
- Personal monitors
- Public transport sensors
- Community networks

Satellite Systems:

Space-based monitoring:

Types of satellites:

- Geostationary
- Polar orbiting
- Specialized missions
- Research platforms
- Weather satellites

Measurement capabilities:

- Aerosol optical depth
- Gas concentrations
- Cloud interactions
- Temperature profiles
- Pollution transport

Data Integration:

Network coordination:

- International standards
- Data sharing protocols
- Quality control
- Calibration procedures
- Reporting systems

Analysis platforms:

- Global databases
- Research networks
- Public interfaces
- Policy support
- Health monitoring

The future of air quality monitoring is looking pretty exciting, with innovative technologies emerging all the time:

Emerging Technologies:

Low-cost sensors:

Miniaturization:

- Chip-based sensors
- Wearable devices
- Smart home integration
- Vehicle sensors
- Building monitors

Network capabilities:

- Wireless communication
- Real-time data
- Cloud integration
- Automated alerts
- Pattern recognition

Artificial Intelligence Applications:

Machine learning:

Pattern detection:

- Anomaly identification
- Trend prediction
- Source attribution
- Risk assessment
- Health forecasting

Data processing:

- Automated analysis
- Quality control
- Gap filling
- Calibration
- Validation

Smart City Integration:

Urban monitoring:

Infrastructure:

- Streetlight sensors
- Traffic monitoring
- Building integration
- Public spaces
- Transportation hubs

Data management:

- Real-time reporting
- Public alerts
- Emergency response
- Planning support
- Policy guidance

Practical Applications:

Personal Use:

Daily planning:

Activity scheduling:

- Outdoor exercise
- Travel timing
- Ventilation decisions
- Protection measures

- Emergency preparation

Health management:

- Symptom tracking
- Exposure reduction
- Medical coordination
- Risk awareness
- Prevention strategies

Community Action:

Local initiatives:

Monitoring programs:

- Citizen science
- School monitoring
- Community networks
- Hot spot identification
- Action planning

Response strategies:

- Public education
- Alert systems
- Protection measures
- Policy advocacy
- Resource sharing

Professional Applications:

Environmental management:

Compliance monitoring:

- Regulatory requirements
- Emission controls
- Impact assessment
- Mitigation planning
- Performance tracking

Research support:

- Health studies
- Environmental impact
- Policy evaluation
- Technology assessment
- Trend analysis

Looking ahead, we can expect air quality monitoring to become even more integrated into our daily lives. **Imagine:**

- Smart homes that automatically adjust ventilation based on both indoor and outdoor air quality
- Wearable devices that provide personalized exposure alerts and health recommendations
- City-wide networks that optimize traffic flow to reduce pollution hot spots
- Predictive systems that warn of potential air quality issues days in advance

- Integration with healthcare systems for better understanding of health impacts

But remember, all this technology is just a tool. The real power comes from how we use this information to protect ourselves and improve our environment. It's like having a sophisticated security system – it's great to have the alerts, but you need to know what to do when they go off.

As we wrap up this chapter, here are some key takeaways for making the most of air quality monitoring:

Practical Tips:

Regular checking:

- Morning planning
- Activity adjustment
- Travel preparation
- Protection measures
- Emergency readiness

Data use:

- Multiple sources
- Local conditions
- Personal sensitivity
- Time patterns
- Action thresholds

Remember, monitoring and assessment are just the beginning. In our next chapter, we'll explore specific actions we can take to protect ourselves and improve air quality. But for now, take a moment to check your local air quality – because knowledge isn't just power, it's protection.

And hey, while those old mood rings might not have been very scientific, at least they got us thinking about monitoring our environment. Today's air quality monitoring systems might not be as fashionable, but they're definitely more useful for keeping us safe and healthy!

CHAPTER 10

PROTECTION AND PREVENTION

Alright, this is it – the chapter you've been waiting for. We've talked about all the problems with our air quality, but now it's time for solutions. Think of this as your personal air quality survival guide, packed with practical steps you can take today to breathe easier tomorrow.

Let's start with personal protection strategies, because let's face it – while we work on fixing the bigger problems, we need to protect ourselves and our loved ones right now. It's like wearing a seatbelt while also supporting better car safety technology – you don't skip the seatbelt just because cars are getting safer.

First up, personal protection strategies that actually work:

Masks and Respirators:

- N95 masks for particle pollution
- P100 respirators for more serious exposure
- Proper fit is crucial (that beard might have to go!)
- Replace regularly
- Know their limitations

Timing Your Activities:

- Check air quality before outdoor exercise
- Avoid peak pollution hours
- Plan indoor workouts on bad air days
- Take alternate routes away from heavy traffic
- Stay indoors during pollution alerts

Personal Habits:

- Shower after heavy exposure
- Change clothes that have been exposed
- Keep windows closed during peak pollution
- Use air conditioning with good filters
- Stay well-hydrated

But personal protection is just the start. Your home should be your clean air sanctuary, not another source of pollution. Here's how to make that happen:

Ventilation Improvements:

- Install high-quality air filters
- Maintain your HVAC system
- Use exhaust fans properly
- Consider whole-house ventilation
- Create positive pressure zones

Air Cleaning Solutions:

- HEPA air purifiers in key rooms

- UV air sanitizers where needed
- Regular cleaning and maintenance
- Proper sizing for your space
- Strategic placement

Source Control:

- Remove shoes at the door
- Ban smoking indoors
- Choose low-VOC products
- Control humidity
- Fix leaks promptly

Think of your home like a submarine – you want to control what comes in and goes out. But instead of keeping water out, you're managing air quality. Here are some room-specific strategies:

Bedroom:

- No electronics charging overnight
- Washable allergen-proof covers
- Regular bedding washing
- Air purifier running
- Good ventilation

Kitchen:

- Strong exhaust fan
- Clean cooking practices

- Regular appliance maintenance
- Natural cleaning products
- Good ventilation

Bathroom:

- Control moisture
- Clean regularly
- Fix leaks promptly
- Use exhaust fan
- Choose safe cleaning products

But we can't just focus on our own homes – we need community-level solutions too. After all, air pollution doesn't stop at your property line. Here's how communities can work together:

Neighborhood Initiatives:

- Car-free zones
- Tree planting programs
- Community air monitoring
- Clean energy projects
- Anti-idling campaigns

Public Spaces:

- Green spaces and parks
- Better public transportation
- Bike lanes and walkways

- Clean air shelters
- Air quality alerts

Community Education:

- School programs
- Public workshops
- Information sharing
- Citizen science projects
- Emergency response plans

Think of your community like a big extended family – when it comes to air quality, we're all in this together. That's why policy and regulation are so important. Here's what effective air quality policies look like:

Environmental Regulations:

- Emission standards
- Industrial controls
- Vehicle inspections
- Construction dust control
- Indoor air quality standards

Enforcement Mechanisms:

- Regular inspections
- Meaningful penalties
- Public reporting
- Transparent monitoring

- Quick response to violations

Economic Incentives:

- Clean energy tax breaks
- Electric vehicle subsidies
- Green building credits
- Pollution reduction rewards
- Research funding

But the really exciting part is what's coming in the future. Modern technologies are emerging that could revolutionize how we deal with air pollution:

Air Cleaning Technologies:

- Artificial trees that absorb CO2
- Self-cleaning buildings
- Pollution-eating materials
- Smog-capturing towers
- Nano-tech air filters

Monitoring Advances:

- Real-time personal monitors
- Satellite tracking
- AI prediction systems
- Micro-sensors everywhere
- Integrated warning systems

Green Technologies:

- Zero-emission vehicles
- Carbon capture systems
- Clean energy storage
- Smart building systems
- Pollution prevention materials

Think of these technologies like reinforcements coming over the horizon – help is on the way, but we need to hold the line until it arrives. That means taking action now with what we have while supporting development of better solutions.

Here's what you can do today:

Personal Actions:

- Check daily air quality
- Use appropriate protection
- Create clean air spaces
- Reduce your pollution footprint
- Stay informed and prepared

Home Improvements:

- Upgrade air filters
- Install air purifiers
- Control pollution sources
- Maintain ventilation systems
- Monitor indoor air quality

Community Involvement:

- Join local initiatives
- Support clean air policies
- Share information
- Report violations
- Help vulnerable neighbors

And now for the call to action – because knowing isn't enough, we need to do something with this knowledge:

First, commit to protecting yourself and your family:

- Get proper protective equipment
- Create your clean air action plan
- Make your home a clean air sanctuary
- Know when to take extra precautions
- Stay informed about air quality

Second, work on reducing your contribution to air pollution:

- Drive less
- Save energy
- Choose clean products
- Maintain your vehicles
- Support clean energy

Third, get involved in your community:

- Support clean air initiatives
- Contact your representatives

- Share what you've learned
- Help vulnerable neighbors
- Join citizen science projects

Fourth, think long-term:

- Support research and development
- Vote for clean air policies
- Invest in clean technologies
- Plan for a cleaner future
- Teach the next generation

Remember, every breath matters. Each action you take to protect yourself and improve air quality ripples out to affect others. It's like throwing a pebble in a pond – the ripples spread far beyond the initial splash.

Don't get overwhelmed by the challenge. Start with one thing:

- Buy an air purifier
- Check daily air quality
- Change your furnace filter
- Join a community group
- Write to your representatives

Then add another. And another. Before you know it, you're making a real difference.

Think of it this way: future generations will either thank us for taking action or wonder why we didn't. Which would you prefer?

The solutions are there. The technologies are improving. The knowledge is spreading. What we need now is action – from individuals, communities, businesses, and governments.

You've read this book. You know the problems. You understand the solutions. Now it's time to act. Because clean air isn't just a nice-to-have – it's essential for life itself.

So, take a deep breath (hopefully of clean air) and get started. Whether it's protecting yourself, improving your home's air quality, joining community initiatives, or supporting better policies, every action counts.

Remember: The best time to start protecting your air was yesterday. The second-best time is now.

CHAPTER 11

ESSENTIAL AIR QUALITY TIPS AND FEATURES

Throughout this book, we've packed in extra bits of wisdom, warnings, and real-world advice to help you navigate the complex world of air quality. Think of this chapter as your Swiss Army knife of air quality information – packed with tools you can use every day to breathe easier.

DID YOU KNOW?

Air pollution can make you gain weight! Recent studies show that certain air pollutants interfere with your metabolism and can contribute to obesity. It's like your body's engine running less efficiently because of dirty fuel.

Let's start with some essential warnings that could save your life or protect your health. These are the kinds of alerts you should never ignore:

WARNING ALERT

Never mix bleach with ammonia or other cleaners. This combination can create toxic gases that can cause severe injury or death within minutes. Always read labels and never mix cleaning products.

Here's what expert pulmonologist Dr. Sarah Chen has to say about protecting your lungs:

"Think of your lungs like a sophisticated filter system. Just as you wouldn't pour sand into your car's gas tank, you shouldn't expose your lungs to harmful pollutants. The damage can be cumulative and irreversible."

💡 QUICK TIP

Keep indoor humidity between 30-50% to prevent mold growth. Use a dehumidifier in damp areas and fix water leaks promptly.

Let's look at a real-world example of how air quality affects daily life:

CASE STUDY

The Johnson Family the Johnsons lived near a busy highway for years, dealing with constant vehicle exhaust. After their youngest developed asthma, they:

- Installed HEPA filters in every bedroom
- Created a clean air room for high pollution days
- Used air quality apps to plan outdoor activities
- Planted a tree barrier along their property

Result: Their child's asthma attacks decreased by 60%

📊 DATA SNAPSHOT

Indoor vs. Outdoor Air Pollution

- Indoor air: Often 2-5x more polluted
- Common indoor pollutants: 50+ chemicals
- Time spent indoors: 90% of our day
- Cleaning products used annually: 25+ pounds per household

Let's break down some action steps you can take right now:

✓ ACTION CHECKLIST

- ☐ Check your home's air filters
- ☐ Download an air quality app
- ☐ Create an air quality emergency kit
- ☐ Know your nearest clean air shelter
- ☐ Have an evacuation plan for severe air quality events

📚 DID YOU KNOW?

Plants aren't great air purifiers! Despite widespread belief, you'd need hundreds of plants to equal one good air purifier. But they're still nice to have around!

Here's another expert weighing in on indoor air quality:

"Most people think outdoor air pollution is worse than indoor air pollution. That's usually not true. Your indoor air is like a concentrated soup of chemicals from cleaning products, furniture, and building materials." - Dr. Michael Torres, Indoor Air Quality Specialist

QUICK REFERENCE GUIDE

Air Quality Index (AQI)

0-50: Great! Enjoy outdoor activities

51-100: Okay for most people

101-150: Sensitive groups should take care

151-200: Everyone should limit outdoor activity

201-300: Serious health effects possible

300+: Hazardous conditions

⚠ WARNING ALERT

Carbon monoxide is odorless and deadly. Never run generators, grills, or cars in enclosed spaces, including garages. Install CO detectors on every floor of your home.

Let's look at another real-world example:

CASE STUDY

Portland School District When wildfires threatened air quality, the district:

- Installed air quality monitors in every school
- Created clean air shelters in gymnasiums
- Developed activity guidelines based on AQI
- Trained staff on emergency procedures

Result: Protected 50,000 students during multiple fire seasons

💡 QUICK TIP

Change your car's cabin air filter every 15,000-25,000 miles. A clean filter can reduce in-car pollution by up to 70%.

Here's what children's health expert Dr. Lisa Wong has to say:

"Children's lungs are still developing until their late teens. Exposure to air pollution during these critical years can have lifelong impacts. Protection during childhood is crucial."

SEASONAL ACTION GUIDE

Spring:

- Check pollen forecasts
- Clean air conditioner filters
- Service HVAC systems
- Plan garden with low-allergen plants

Summer:

- Monitor for wildfire smoke
- Use early morning hours for exercise
- Keep windows closed on high ozone days
- Check air quality before outdoor events

Fall:

- Clean gutters to prevent mold
- Check heating system filters
- Seal drafts and leaks
- Create winter emergency kit

Winter:

- Monitor indoor humidity
- Check fireplace and wood stove safety
- Keep paths clear for ventilation
- Test carbon monoxide detectors

📚 DID YOU KNOW?

Your car's air is typically worse than outside air! Cars trap and concentrate pollution from surrounding traffic. Use your recirculate button in heavy traffic.

EMERGENCY KIT CHECKLIST

- ☐ N95 masks (one per family member)
- ☐ Portable air purifier
- ☐ Battery-powered radio
- ☐ Extra medications
- ☐ Emergency contact numbers
- ☐ Evacuation plan
- ☐ Clean air shelter locations

⚠️ WARNING ALERT

Dust from home renovation can contain lead, asbestos, and other toxic materials. Always test before disturbing old materials and use proper protection.

Here's insight from environmental justice advocate Maria Rodriguez:

"Poor air quality doesn't affect everyone equally. Low-income communities often face the worst pollution but have the fewest resources to deal with it. We need solutions that work for everyone."

QUICK POLLUTION SOLUTIONS

Immediate: Close windows during peak traffic
Short-term: Use air purifiers
Medium-term: Improve home ventilation
Long-term: Support clean air policies

💡 QUICK TIP

Your vacuum could be making air quality worse! Use one with a HEPA filter and empty it outside to prevent dust from recycling through your home.

Let's examine one more real-world example:

CASE STUDY

City of Phoenix During extreme heat and pollution events, Phoenix:

- Created public clean air centers
- Provided free filtration to vulnerable residents
- Implemented text alert system
- Coordinated with healthcare providers

Result: 30% reduction in pollution-related ER visits

COMMUNICATION GUIDE

When to Contact:

Doctor: Persistent respiratory symptoms
Air Quality Agency: Visible pollution sources
Emergency Services: Immediate health threats
Local Government: Community air quality concerns

📚 DID YOU KNOW?

Opening windows doesn't always improve air quality! Sometimes it makes it worse. Check outdoor air quality before ventilating.

Remember these key action points:

Monitor: Keep track of air quality daily
Protect: Use appropriate protection when needed
Prevent: Reduce pollution sources in your control

Prepare: Have emergency plans ready

Share: Help others understand air quality risks

The tools and information in this chapter are your air quality Swiss Army knife – keep them handy and use them often. Because when it comes to air quality, knowledge isn't just power – it's protection.

CHAPTER 12

KEY HIGHLIGHTS AND READER TAKEAWAYS

Understanding Air Quality

- The composition of clean air vs. polluted air
- Natural and artificial pollutants.
- How air quality affects different populations differently

Indoor Air Dangers

- Why indoor air is often more polluted than outdoor air
- Common household pollutant sources
- How to create cleaner air spaces
- Health Impacts
- Short-term and long-term health effects
- Vulnerable populations
- Warning signs and symptoms

Protection Strategies

- Personal protection equipment
- Home air quality improvements
- Community-level solutions

Monitoring and Assessment

- How to read air quality indexes
- Personal monitoring tools
- When to take action

Future Outlook

- Emerging technologies
- Policy directions
- Individual and community actions

Closing Message to Readers

Thank you for joining me on this journey through the invisible world of air quality. Understanding what we breathe is just the first step – now it's time for action. Whether you start with a single air purifier or join a community clean air initiative, every step toward better air quality counts. Remember, we all share the same air, and together we can make every breath a healthier one.

Share what you've learned from this book with others. Keep this guide handy as a reference, and most importantly, stay aware of your air. Your lungs will thank you, and so will future generations. Here's to breathing easier, one informed decision at a time.

Printed in the USA
CPSIA information can be obtained
at www.ICGtesting.com
CBHW051743101224
18757CB00010B/502